Ecological Research Monographs

Asia-Pacific Biodiversity Observation Network

Series editor
Yoh Iwasa

More information about this series at http://www.springer.com/series/8852

Shin-ichi Nakano • Tetsukazu Yahara •
Tohru Nakashizuka
Editors

Aquatic Biodiversity Conservation and Ecosystem Services

Springer

Editors
Shin-ichi Nakano
Center for Ecological Research
Kyoto University
Otsu, Shiga
Japan

Tetsukazu Yahara
Department of Biology
Kyushu University
Fukuoka, Japan

Tohru Nakashizuka
Graduate School of Life Sciences
Tohoku University
Sendai, Japan

ISSN 2191-0707 ISSN 2191-0715 (electronic)
Ecological Research Monographs
ISBN 978-981-10-0778-1 ISBN 978-981-10-0780-4 (eBook)
DOI 10.1007/978-981-10-0780-4

Library of Congress Control Number: 2016939121

Front cover: Plecoglossus altivelis, sweetfish (*ayu*), swimming near the shore of Lake Biwa, Japan (photo by Kohta Sawada). *Back cover*: *Left*: Chinese cuisine prepared on a boat in Lake Taihu, China. The ingredients are an ecosystem service of the lake (photo by Shin-ichi Nakano). *Center*: *Nypa fruticans* Wurmb, nipa palm, in a tributary of the Mekong River, Vietnam (photo by Shin-ichi Nakano). *Right*: Phytoplankton assemblage of Lake Biwa, Japan (photo by Shohei Fujinaga).

Printed on acid-free paper

This Springer imprint is published by Springer Nature
The registered company is Springer Science+Business Media Singapore Pte Ltd.

Preface

It has been commonly accepted that biological diversity is important as a natural resource and also for functioning of the ecosystem which provides us with ecosystem services essential for human well-being, although biodiversity loss is continuing both on regional and global scales. To provide a quantum step forward in advancing science that optimizes the synergy between development and biodiversity conservation in Asia, we have published two volumes from the Asia-Pacific Biodiversity Observation Network (AP-BON): *The Biodiversity Observation Network in the Asia-Pacific Region: Toward Further Development of Monitoring* and *Integrative Observations and Assessments*. In those two books, we have mainly introduced the status quo of biodiversity monitoring in Asia, together with advanced concepts and methods for biodiversity and ecosystem service.

In October 2014, the 12th meeting of the Conference of the Parties to the Convention on Biological Diversity (CBD-COP12) was held in Pyeongchang, South Korea. According to the mid-term review of progress toward the goals of the Strategic Plan for Biodiversity 2011–2020 and the Aichi Biodiversity Targets using the fourth edition of the Global Biodiversity Outlook (GBO4) as background material, it was recognized that we had made progress toward meeting some elements of most Aichi Biodiversity Targets. However, in most cases, these advances will not be enough to achieve the targets, and further urgent and effective action is needed to reduce the threats to biodiversity and to prevent its continued decline. At CBD-COP12, the "Gangwon Declaration", the report from the ministerial level meeting where sustainable development with special reference to biodiversity was discussed, was adopted, and the significant contribution of biodiversity due to its critical foundation of the Earth system on which human well-being depends is clearly stated.

In November 2012, we held the International Workshop on Freshwater Biodiversity Conservation in Asia at Kyushu University, Japan, to discuss how to identify the biodiversity hotspots, appropriate methods to identify the drivers of biodiversity loss of particular freshwater systems, and data sharing among

Asian countries. The chapters drawn from the workshop presentations dominate in this third volume from AP-BON, referring to local biodiversity, management and conservation of biodiversity, and newly developed concepts and methodologies in freshwater systems. In addition, we also have invited some chapters on a marine system and ecosystem service and socioeconomic aspects with special reference to biodiversity because of the analogy to freshwater ecosystems.

The first part, "Local Biodiversity and Its Threats", consists of four chapters. Our planet is currently facing a freshwater biodiversity crisis, and the key to preventing further extinction is found by understanding the threats facing aquatic habitats.

Better management of biodiversity conservation requires frequent and spatially detailed assessments of species numbers and distributions. To collect such information, laborious and expensive support is needed. The usefulness of modern sophisticated technologies that measure the distribution and status of biodiversity is probably an ideal way to gather these crucial data. Thus in the second part, "Advanced Methods of Biodiversity Monitoring", we include a chapter on remote sensing technology applied to a eutrophic lake and another chapter from marine seagrass beds.

Ecosystem service in terms of biodiversity is now attracting increasing interest all over the world. When an ecosystem provides services to humans, some interaction with humans is required. Therefore, socioeconomic aspects with special reference to biodiversity would be an appropriate approach to evaluate ecosystem services in a given ecosystem. In the third part, "Ecosystem Service and Socioeconomic Aspects with Special Reference to Biodiversity", we have included one chapter on Indonesian freshwater fisheries and, to provide a broader perspective, two more chapters on terrestrial ecosystem services. The present book, together with our previous books, presents the status quo of Asian biodiversity in the biodiversity research that still lacks information from developing countries. In addition, we have included contributions, providing reviews on advances in concepts and methods of biodiversity observations and on the challenges to study spatial variability of biodiversity and ecosystems by linking monitoring across various ecosystems in the Asia-Pacific region. These contributions are important for effective conservation and sustainable use of aquatic biodiversity in this region.

We hope the present book will be informative for all the stakeholders interested in biodiversity issues: researchers, policy (decision) makers, NPOs, NGOs, and industries related to environmental issues.

We are very grateful to the Ministry of Environment, Japan, for providing administrative and financial support. Also, we would like to thank the authors for submitting their manuscripts; to the Secretariat of DIVERSITAS of the Western Pacific and Asia (DIWPA) for its formatting submitted manuscripts; and to the publisher Springer for its patience with our delayed editing of the book.

Otsu, Japan — Shin-ichi Nakano
Fukuoka, Japan — Tetsukazu Yahara
Sendai, Japan — Tohru Nakashizuka

Contents

Contributors

Zulkifli Dahlan Environmental Doctoral Programme, Sriwijaya University, South Sumatra, Indonesia

Masahiko Fujii Faculty of Environmental Earth Science, Hokkaido University, Hokkaido, Japan

Takehiko Fukushima Faculty of Life and Environmental Sciences, University of Tsukuba, Ibaraki, Japan

Abdul Karim Gaffar Research Institute for Inland Fisheries, South Sumatra, Indonesia

Nicolas Gruber Environmental Physics Group, Institute of Biogeochemistry and Pollutant Dynamics, Zurich, Switzerland

Chaiwut Grudpan Faculty of Agriculture, Ubon Ratchathani University, Ubon Ratchathani, Thailand

Liangliang Huang College of Environmental Science and Engineering, Guilin University of Technology, Guilin, China

Satoshi Ishiguro National Institute for Environmental Studies, Ibaraki, Japan

Tuantong Jutagate Faculty of Agriculture, Ubon Ratchathani University, Ubon Ratchathani, Thailand

Ryo Kohsaka Graduate School of Human and Socio-Enviromental Studies, Kanazawa University, Ishikawa, Japan

Riky Kurniawan Research Center for Limnology, Indonesian Institute of Sciences, Cibinong, Indonesia

Jianhua Li College of Environmental Science and Engineering, Tongji University, Shanghai, China

Tsuneo Matsunaga National Institute for Environmental Studies, Ibaraki, Japan

Bunkei Matsushita Faculty of Life and Environmental Sciences, University of Tsukuba, Ibaraki, Japan

Hikaru Matuoka Kanazawa University Institute of Science and Engineering, Ishikawa, Japan

Dina Muthmainnah Research Institute for Inland Fisheries, South Sumatra, Indonesia

Kaho Naganawa School of Rural Development, Kanazawa University, Ishikawa, Japan

Hiroyuki Oguma National Institute for Environmental Studies, Ibaraki, Japan

Yoichi Oyama Marimo Research Center, Kushiro Board of Education, Hokkaido, Japan

Dwi Putro Priadi Environmental Doctoral Programme, Sriwijaya University, South Sumatra, Indonesia

Yasushi Shoji Research Faculty of Agriculture, Hokkaido University, Hokkaido, Japan

Marco Steinacher Climate and Environmental Physics, Physics Institute, University of Bern, Bern, Switzerland

Luki Subehi Research Centre for Limnology, Indonesian Institute of Sciences, Cibinong, Indonesia

Robiyanto H. Susanto Environmental Doctoral Programme, Sriwijaya University, South Sumatra, Indonesia

Apinun Suvarnaraksha Faculty of Fisheries Technology and Aquatic Resources, Maejo University, Chiangmai, Thailand

Mitsuyuki Tomiyoshi Graduate School of Human and Socio-Environmental Studies, Kanazawa University, Ishikawa, Japan

Triyanto Research Centre for Limnology, Indonesian Institute of Sciences, Cibinong, Indonesia

Meike Vogt Environmental Physics Group, Institute of Biogeochemistry and Pollutant Dynamics, Zurich, Switzerland

Katsumasa Yamada Research Center for Fisheries and Environment in the Ariake and Yatsushiro Bay, Seikai National Fisheries Research Institute, Fisheries Research Agency, Nagasaki, Japan

Takehisa Yamakita Japan Agency for Marine Earth, Science and Technology, Institute of Biogeosciences, Kanagawa, Japan

Yasuhiro Yamanaka Faculty of Environmental Earth Science, Hokkaido University, Hokkaido, Japan

Hiroya Yamano Center for Environmental Biology and Ecosystem Studies, National Institute for Environmental Studies, Ibaraki, Japan

Yumiko Yara Center for Environmental Biology and Ecosystem Studies, National Institute for Environmental Studies, Ibaraki, Japan

Project Team for Analyses of Changes in East Japan Marine Ecosystems, Japan Agency for Marine-Earth Science and Technology (JAMSTEC), Kanagawa, Japan

Part I
Local Biodiversity and Its Threats

Chapter 1
Aquatic Macrophytes and Fish Diversity of Various Tropical Lakes at the Main Islands in Indonesia

Riky Kurniawan, Triyanto, and Luki Subehi

Abstract Indonesia as a tropical country has unique characteristics and is one of the biggest countries for listed freshwater macrophytes and fish species in Asia. As for freshwater, Indonesian lakes are comprised of 521 natural lakes and over 100 reservoirs, which cover about 21,000 km2 of area and impound approximately 500 km3 of water. The objective of this study is to collect information about aquatic macrophytes and fish diversity conditions. As one of the megadiverse countries, Indonesia is acclaimed that 10 % of plants occurring in various habitats are established steadily in this country. Moreover, those plants are scattered and isolated in the islands for a long time; hence, it results in an ultimately high level of endemicity. Ecologically, they provide shelters and spawning sites for fish and work as purification system, water current-stabilizing system, and erosion and turbidity controller. The results showed that most spread species in the identified lakes are *Eicchornia crassipes* (water hyacinth) and *Cyprinus carpio*. However, more researches are needed to estimate the current state of aquatic plant and fish diversity. In spite of limitations of the research for the aquatic plant and fish diversity, the opportunity to explore its diversity is still widely opened.

Keywords Macrophytes • Fish • Diversity • Tropical • Lakes • Aquatic • Ecologically

Introduction

Indonesia as a tropical country has unique characteristics. It is listed as 1 of the 12 megadiverse countries, ranked second place for the number of freshwater fish species, and reknowned as one of the biggest countries for listed freshwater fish species in Asia. As for freshwater, Indonesian lakes are comprised of 521 natural lakes and over 100 reservoirs, which cover about 21,000 km^2 of area and impound

R. Kurniawan • Triyanto • L. Subehi (✉)
Research Center for Limnology, Indonesian Institute of Sciences, Cibinong 16911, Indonesia
e-mail: luki@limnologi.lipi.go.id

S.-i. Nakano et al. (eds.), *Aquatic Biodiversity Conservation and Ecosystem Services*, Ecological Research Monographs, DOI 10.1007/978-981-10-0780-4_1

approximately 500 km^3 of water. It is used for both ecological and economic services.

As endorsed by the National Conference for Lakes in Indonesia II, held on October 13–14, 2011, in Semarang, the Ministry of National Environment restated that 15 lakes are nationally treasured as the priority on environmental crisis to resolve. The assessment criteria for designated priority lakes (2010–2014) are as follows:

- Lake damages: the level of sedimentation, pollution, eutrophication, highly reducing water quality, and quantity of water
- Utilization of lakes: hydropower plants, agriculture, fisheries (aquaculture/floating cage), usable water, religious and cultural values, tourism (including uniqueness of lakes, accessibility, amenities, infrastructures, and conditions of society)
- Local government's and society's commitments to wise management of the lakes (master plans, local regulations (*perda*), and managing committees)
- Strategic lakes: lakes featuring strategic functions of national interests
- Biodiversity (including endemic fish species, aves, and vegetation)
- Carbon urgency (the challenge to global climate change)

In addition, the problem of biodiversity crisis in limnetic systems is worse than any other natural systems (Millennium Ecosystem Assessment 2005). To identify the reason for the crisis in limnetic environments and to conduct better conservation and managements, we must urgently collect information about the status quo of limnetic through appropriate environmental monitoring (Williams et al. 2002).

Activities and Issues of Freshwater Biodiversity Monitoring

As one of megadiverse countries, Indonesia is acclaimed that 10 % of plants occurring in various habitats are established steadily in this country (Mahyar and Sadili 2003; LBN 1977). Moreover, those plants are scattered and isolated in the islands for a long time; hence, it results in an ultimately high level of endemicity. Meanwhile, in terms of species richness, this country is blessed with abundant species in the mixture of two biogeographical regions (i.e., Oriental and Australian) (Whitten 1994). The condition makes Indonesia one of the most spotlighted countries for worldwide biodiversity research.

Aquatic macrophytes in Indonesia including the species of microflora (e.g., ferns, mosses) and Aangiospermae are a specific group to be focused on research not only for the diversity but also for the specific functions and services. The presence of controlled aquatic macrophytes in the inland water ecosystem is recognized as one of the most important factors to support the ecosystem sustainability. Ecologically, they provide shelters and spawning sites for fish and work as purification system, water current-stabilizing system, and erosion and turbidity controller. In the same manner, they also assume a role of economic function for people, as food and fodder production (Newall 1995; Wetzel and Gopal 2001).

Ward et al. (1994) highlighted the necessity that aquatic macrophyte diversity has a strong correlation with the diversity of fauna in the ecosystem.

On the one hand, although the imperative of aquatic macrophytes has been recognized in Indonesia, there are only few researches focusing on this specific group. In addition, these researches are somehow old, less updated, scattered, and unorganized. Hence, it makes difficult for the recent researchers to continue on the previous works even to start from other new researches as a completed investigation, especially in aquatic macrophyte field.

Taxonomists on aquatic macrophytes are absolutely rare in Indonesia. In consequence, research on this specific group is almost vanished. Additionally, the current trend on lake development programs tends to the shoreline development of hotels and restaurants on littoral zones. This trend absolutely causes more damages on the zone most plants require as their habitats. Hence, it may be one of the most influencing factors of the loss of aquatic macrophyte diversity.

Even though those constraints may limit research on biodiversity, they may give challenges to the young generation to continue the research since aquatic plants provide actually not only negative effects but also many positive sides. Significantly, it is an obligation for the scientist to get the decision maker in a good management regime aware that aquatic macrophytes support people prosperity.

Indonesia has the highest diversity of freshwater fish, ranked second place in the world followed by Brazil. The diversity of freshwater fish is scattered in various freshwater habitats like rivers, lakes, swamps, peat lands, and brackish waters. According to Kottelat et al. (1993) and Hartoto et al. (2009), there are more than 1,000 species of freshwater fish in Indonesia. Most of Indonesia's fishes have been described in "The Fishes of the Indo-Australian Archipelago," Vol. 11, edited by M. Weber and L. F. de Beaufort between 1911 and 1962.

Various problems faced by the diversity of freshwater fish in Indonesia such as habitat destruction and water pollution are caused by anthropogenic activities and the invasion of alien fishes that threaten the sustainability of fish diversity (Rahardjo 2011).

Meanwhile there are many lakes that have been deteriorated; many others are still pristine. Accordingly, there are possibilities that other natives or even endemic species have not been recognized. It is a challenge that should be considered by the scientists of aquatic plants and fish. Moreover, there may be discrepancies of aquatic community structure among lakes which depend on the diversity of lake formations, histories, and environments. The abundant natural diversity itself has created one big challenge to be discovered by future researches.

Invasive Alien Species in Indonesia

Alien species are brought into an ecosystem unnaturally. Invasive species whether native or not, which widely affect habitat, cause environmental damages, economic losses, and harms to humans. Alien species are not always invasive. Invasive species do not necessarily come from outside. Invasive alien species is a combination of alien species and invasive species (Anonymous; CBD-UNEP 2011).

Most alien species cannot thrive in the introduction site because they do not fit the new environment. However, some of the species can grow and thrive in new locations, and many of them are classified as invasive species or weeds. Through competition for limited resources, invasive species can replace native species or change the conditions of habitat so that the native species cannot survive longer (Indrawan et al. 2007).

The introduction of alien species in Indonesia can occur in several ways: colonization by foreign nations, agricultural and fishery cultivation, unintentional transports, and research activities. According to Rahardjo (2011), the introduction of alien species into waters provides benefits such as food and other products in the waters, but on the other, alien species have negative impacts on native species as competitor for the same ecological niches, such as disrupting food web, reducing biodiversity, threatening native fish populations, reducing the quality of habitat, and destructing commercial fisheries and aquaculture.

Biodiversity in Lakes from the Large Islands in Indonesia

Based on major 15 lakes in Indonesia, we selected seven lakes that represent each island in various types.

Sumatera Island

Lake Toba

Lake Toba is the largest lake in Indonesia and Southeast Asia. It is located in the province of North Sumatra. The lake is located at an altitude of 903 m above sea level and with surface water area of 1.130 km^2. The average depth of Lake Toba is 223 m with a maximum depth of 529 m.

Based on Wetlands International Indonesia (1990), aquatic macrophyte biodiversity in Lake Toba is dominated by floating plants of family Nymphaeaceae (*Nelumbo nucifera* and *Nymphaea pubescens*) and Polygonaceae (*Polygonum barbatum* and *Polygonum pulchrum*) as shown in Fig. 1.1. In particular, a species of *Eicchornia crassipes* (Water hyacinth) is the most dominant species covering the lake's surface area. On the other hand, the biodiversity of fish in Lake Toba is dominated by family Cyprinidae (*Barbodes gonionotus*, *Cyprinus carpio*, *Mystacoleucus padangensis*, *Puntius binotatus*, and *Rasbora jacobsoni*).

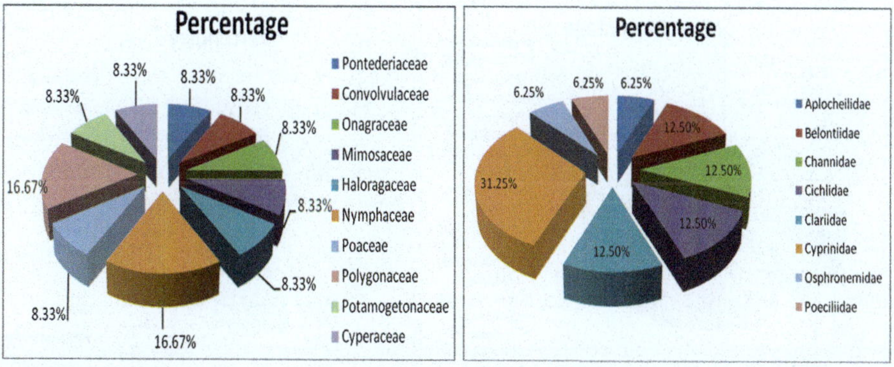

Fig. 1.1 Percentage family of aquatic macrophytes (*left*) and fishes (*right*) in Lake Toba

Lake Kerinci

Lake Kerinci is a volcanic lake with 4.200 ha extents, located in Jambi Province, Sumatera. Geographically, it is located at 2°7′28″–2°8′14″ S and 101°31′34″–101°26′50″ E. It has a surface area of 46 km^2 and 1.6 million m^3 of water volume, and the average depth reaches 97 m.

Biodiversity in Lake Kerinci is reported by Giesen and Sukotjo (1991) and Wetlands International Indonesia (1990). Figure 1.2 shows aquatic macrophyte biodiversity in Lake Kerinci. They are dominated by emergent plants of family Cyperaceae (*Scirpus grossus*, *Cyperus platystylis*, *Fuirena umbellata*, and *Rhynchospora corymbosa*) and Poaceae (*Panicum repens*, *Echinochloa stagnina*, *Leersia hexandra*, and *Hymenachne acutigluma*). In particular, *Eicchornia crassipes* (water hyacinth) is the most dominant species covering the surface of the lake. On the other hand, the biodiversity of fish in Lake Kerinci is dominated by family Cyprinidae (*Barbodes gonionotus*, *Cyprinus carpio*, *Ctenopharyngodon idellus*, *Hampala macrolepidota*, *Osteochilus waandersii*, and *Tor tambroides*).

Jawa: Bali

Lake Batur

Lake Batur is a caldera lake in Bali Province. Geographically, it is located at 8°13′24.0″–8°17′13.3″ S and 115°22′42.3″–115°25′33.0″ E. The surface area of the lake is 16.1 km^2, with a volume of 815.4 million m^3 and an average depth of 50.8 m.

Figure 1.3 shows aquatic macrophyte biodiversity in Lake Batur. They are dominated by floating plants of family Araceae (*Lemna perpusilla* and *Pistia stratiotes*) (Kementerian Lingkungan Hidup Republik Indonesia 2011; Wetlands International Indonesia 1990). In particular, a species of *Eicchornia crassipes*

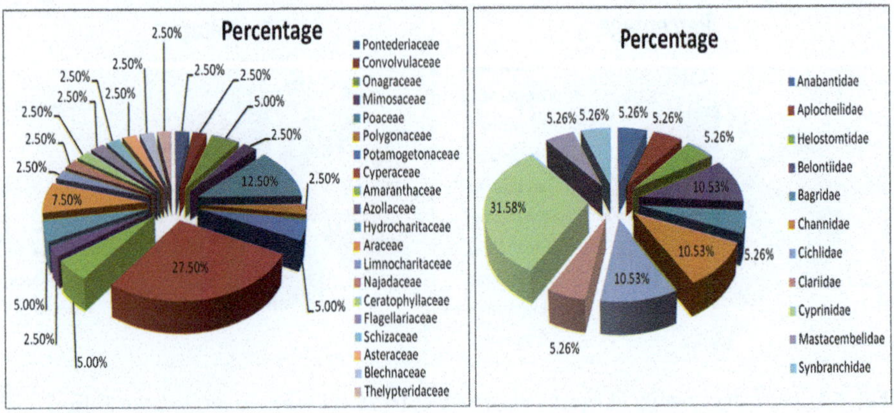

Fig. 1.2 Percentage family of aquatic macrophytes (*left*) and fishes (*right*) in Lake Kerinci

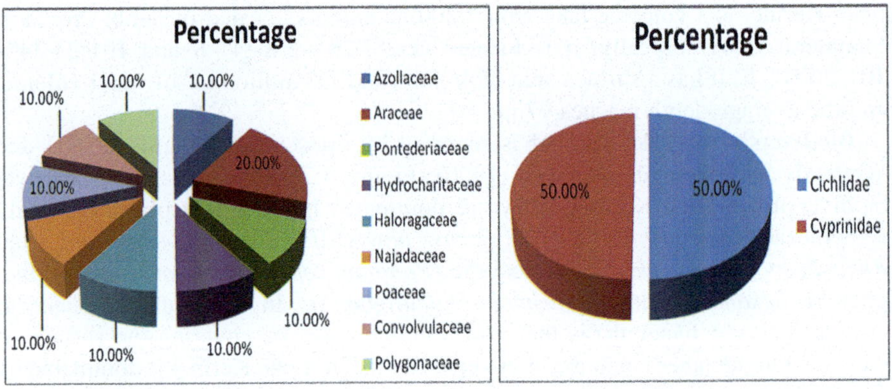

Fig. 1.3 Percentage family of aquatic macrophytes (*left*) and fishes (*right*) in Lake Batur

(water hyacinth) is the most dominant species covering the surface of the lake. In addition, the biodiversity of fish in Lake Batur consists of two families, the Cichlidae (*Oreochromis niloticus* and *Oreochromis mossambicus*) and Cyprinidae (*Cyprinus carpio* and *Rasbora lateristriata*).

Kalimantan

Lake Semayang-Melintang

Lake Semayang-Melintang has an elevation of 17,3 m above sea level. The lake is located in the province of East Kalimantan. This lake is categorized as floodplain

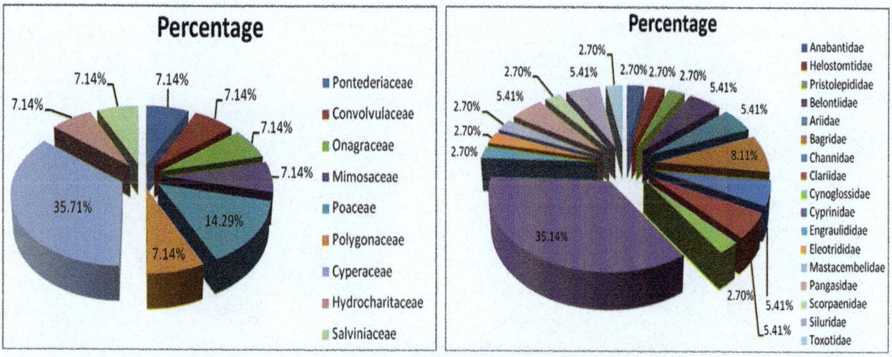

Fig. 1.4 Percentage family of aquatic macrophytes (*left*) and fishes (*right*) in Lake Semayang-Melintang

with muddy and sandy floor. The surface area of Lake Semayang-Melintang is 240 km^2, with a maximum depth of 6.5 m.

Aquatic macrophyte biodiversity at Lake Semayang-Melintang is dominated by emergent plants of family Cyperaceae (*Scirpus grossus*, *Cyperus digitatus*, *Kyllinga brevifolius*, *Cyperus cephalotes*, and *Fimbristylis griffithii*) as shown in Fig. 1.4. In particular, a species of *Eicchornia crassipes* (water hyacinth) is the most dominant species covering the lake's surface area (Nofdianto 1996; Wetlands International Indonesia 1990). In addition, the biodiversity of fish in Lake Semayang-Melintang is dominated by family Cyprinidae (*Puntius lineatus*, *Barbodes schwanenfeldii*, *Barbichtys laevis*, *Chela oxygastroides*, *Osteochilus hasselti*, *Osteochilus wandersi*, and *Osteochilus kelabau*).

Sulawesi

Lake Tempe

Lake Tempe is one of the greatest lakes located in the province of South Sulawesi. Geographically, it is located at 4°00°00″–4°15°00″ S and 119°52°30″–120°07°30″ E. This lake has an area of 47.800 ha. This lake is floodplain lake type with elevation of 10 m above sea level. The depths of the lake during the rainy and dry seasons are 3 m and 1 m, respectively.

Aquatic macrophyte biodiversity in Lake Tempe is dominated by emergent plants of family Poaceae (*Panicum repens* and *Echinochloa crus-galli*) as shown in Fig. 1.5. In particular, Kurniawan (2012) and Wetlands International Indonesia (1990) reported that a species of *Eicchornia crassipes* (water hyacinth) is the most dominant species covering the lake's surface area. On the other hand, the biodiversity of fish in Tempe Lake is dominated by family Cyprinidae (*Barbodes*

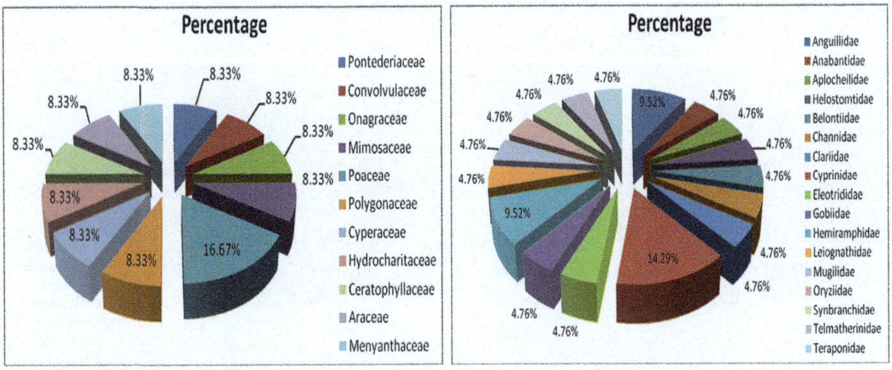

Fig. 1.5 Percentage family of aquatic macrophytes (*left*) and fishes (*right*) in Lake Tempe

gonionotus, *Cyprinus carpio* and *Osteochilus hasselti*) and Anguillidae (*Anguilla marmorata* and *Anguilla nebulosa*).

Lake Matano

Lake Matano is located in the district of East Luwu, province of South Sulawesi, as a tectonic type of lake. Lake Matano has a maximum depth of 595 m with an area of 16.408 ha. Lake Matano, the deepest lake in Southeast Asia, is known as eighth deepest lake in the world.

Figure 1.6 shows aquatic macrophyte biodiversity at Lake Matano (Wetlands International Indonesia 1990). They are dominated by emergent plants of family Cyperaceae (*Cyperus halpan*, *Fimbristylis dichotoma*, *Eleocharis dulcis*, *Eleocharis geniculata*, *Rhynchospora rugosa*, and *Scirpus mucronatus*). In this lake, *Ottelia mesenterium* was an endemic aquatic macrophyte. In addition, the biodiversity of fish in Lake Matano is dominated by family Telmatherinidae (*Telmatherina abendanoni*, *Telmatherina bonti*, *Telmatherina antoniae*, *Telmatherina opudi*, *Telmatherina obscura*, *Telmatherina proghatha*, *Telmatherina sarasinorum*, *Telmatherina wahyui*, and *Telmatherina* sp).

Papua

Lake Sentani

Lake Sentani is located in the district of Jayapura, Papua Province. Geographically, it is located at 140°23′–140°50′ S and 2°31′–2°41′ E. It has an area of 9.630 ha with a maximum depth of 42 m.

Aquatic macrophyte biodiversity in Lake Sentani is dominated by plants of family Araceae (*Lemna perpusilla*, *Spirodela polyrhiza*, and *Pistia stratiotes*) and Hydrocharitaceae (*Hydrilla verticillata* and *Vallisneria natans*) as shown in

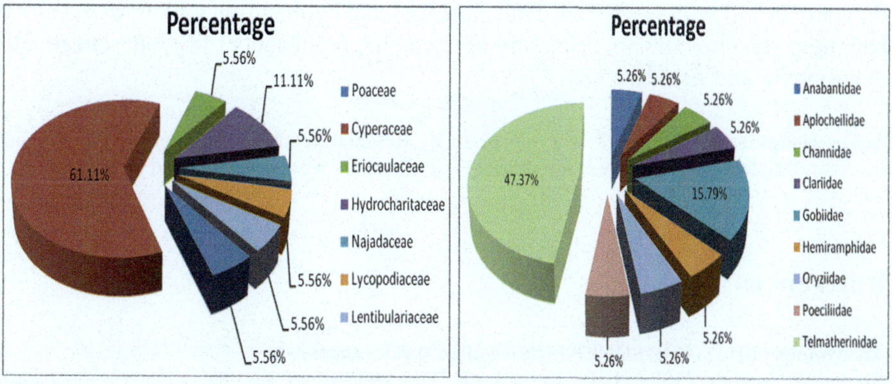

Fig. 1.6 Percentage family of aquatic macrophytes (*left*) and fishes (*right*) in Lake Matano

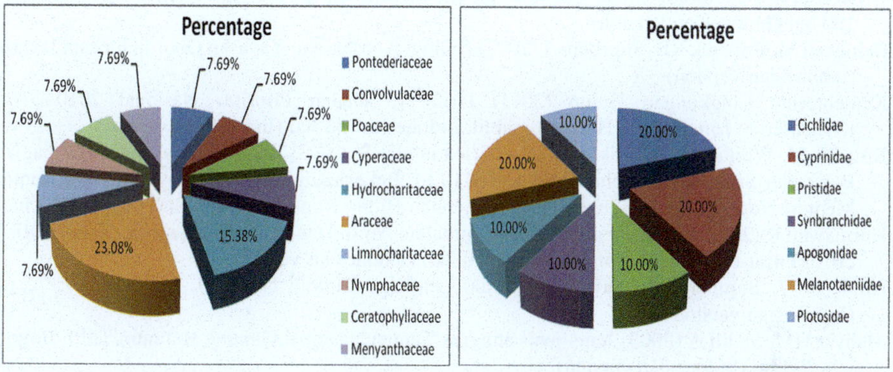

Fig. 1.7 Percentage family of aquatic macrophytes (*left*) and fishes (*right*) in Lake Sentani

Fig. 1.7. In particular, a species of *Eicchornia crassipes* (Pontederiaceae) is the most dominant species covering the lake's surface area (Wetlands International Indonesia 1990). On the other hand, the biodiversity of fish in Lake Sentani is dominated by family Cichlidae (*Oreochromis niloticus*, *Oreochromis mossambicus*), Cyprinidae (*Barbodes gonionotus* and *Cyprinus carpio*), and Melanotaeniidae (*Glossolepis incisus*).

Conclusions

The most spread species in the identified lakes are *Eicchornia crassipes* (Water hyacinth) and *Cyprinus carpio*. However, more researches are needed to estimate the current state of aquatic plant and fish diversity. In spite of limitations of the research for the aquatic plant and fish diversity, the opportunity to explore its diversity is still widely opened. Thus, linkages and cooperation with other scientists and more stakeholders are necessary to reveal the real beauty of Indonesian aquatic macrophyte and fish diversity.

Acknowledgments This work was supported by Research Centre for Limnology, Indonesian Institute of Sciences. We express our thanks to all members who were involved in this project.

References

Anonymous (2011) (CBD)-UNEP invasive alien species Indonesia

Giesen W, Sukotjo (1991) Lake Kerinci and The Wetland of Kerinci Seblat National Park, Sumatra. PHPA/AWB Sumatra Wetland Project Report No. 14. Directorate General of Forest Protection and Nature Conservation, Asian Wetland Bureau-Indonesia

Hartoto DI, dan Nasution SH, Kamal MM (2009) Pedoman Identifikasi Lokasi Calon kawasan Konservasi Perairan Umum Daratan. Direktorat KTNL, Direktorat Jenderal KP33K, KKP. 190 hal (Indonesian version)

Indrawan M, Primack RB, Supriatna J (2007) Biologi Konservasi. Yayasan Obor Indonesia Jakarta (Indonesian version)

Kementerian Lingkungan Hidup (2011) Profil 15 Danau Prioritas Nasional 2010–2014. Kementerian Lingkungan Hidup Republik Indonesia. 148 hal (Indonesian version)

Kottelat M, Whitten AJ, Kartikasari SN, Wirjoatmodjo S (1993) Ikan air tawar Indonesia bagian Barat dan Sulawesi. Periplus Edition (HK) Ltd. Bekerjasama dengan Proyek EMDI, Kantor Menteri Negara Kependudukan dan Lingkungan Hidup Republik Indonesia. Jakarta. 293 hal

Kurniawan R (2012) Penyebaran Populasi Tumbuhan Air di Danau Tempe. Laporan Teknis DIPA 2012. Pusat Penelitian Limnologi LIPI. 10 hal. (Indonesian version)

LBN (1977) Sumber daya hayati Indonesia. Lembaga Biologi Nasional -LIPI, Bogor: 72 pp (Indonesian version)

Mahyar UM, Sadili A (2003) Jenis-jenis anggrek Taman Nasional Gunung Halimun. LIPI, Bogor: ii + 208 pp (Indonesian version)

Millennium Ecosystem Assessment (2005) Ecosystems and human well-being: synthesis. Island Press, Washington, DC

Newall AM (1995) The microflow environment of aquatic plants – an ecological perspective. In: Harper M, Ferguson JD (eds) The ecological basis for river management. Wiley, Chicester

Nofdianto (1996) Ekotipologi dan Fungsional Tumbuhan Air di Perairan Semayang, Kalimantan Timur. Dalam: Rehabilitasi Lingkungan Perairan Danau Semayang, Kalimantan Timur. Puslitbang Ekonomi dan Pembangunan-LIPI. Jakarta. Hal 33–44

Rahardjo MF (2011) Spesies Akuatik Asing Invasif. Prosiding Forum Nasional Pemacuan Sumber Daya Ikan III, Kementerian Kelautan dan Perikanan, Badan Penelitian dan Pengembangan Kelautan dan Perikanan, Pusat Penelitian Pengelolaan Perikanan dan Konservasi Sumberdaya Ikan, Balai Penelitian Pemulihan dan Konservasi Sumberdaya Ikan. Hal KSI-31:1–7 (Indonesian version)

Ward D, Holmes N, Jose P (1994) The new river & wildlife handbook. Royal Society for the Protection of Birds

Wetlands International Indonesia (1990) Accesed 15 August 2012. http://wetlands.or.id/wdb/siteinfo.php?SITE_COD=SUL18

Wetzel RG, Gopal B (2001) Limnology in developing countries. International Association of Theoretical and Applied Limnology and International Scientific Publications, New Delhi

Whitten T (1994) Indonesian heritage 4: Plants. Periplus editions, Singapore: 144 pp

Williams PH, Margules CR, Hilbert DW (2002) Data requirements and data sources for biodiversity priority area selection. J. Biosci 27(Suppl. 2):327–338

Chapter 2
Status of Freshwater Fish Biodiversity in the Yangtze River Basin, China

Liangliang Huang and Jianhua Li

Abstract Chinese freshwater fish numbers are over 1000 species, and at least 717 species in 33 families inhabit rivers, and another 66 species spend part of their lives in rivers, while others are mainly confined to estuarine reaches but occasionally swim upstream. Information on freshwater fish biodiversity in the Yangtze River basin was synthesized. Three hundred sixty-one species were found in the basin, of which 177 species are endemic to the Yangtze River and 25 species are categorized as endangered on the *China Red Data Book* for fishes. Around 80 % of the species (including 124 endemics) in the Yangtze River occur in its upper course above the Three Gorges Dam. As to species richness patterns along the altitude gradient, non-endemic and endemic fishes were different. Non-endemic richness showed a significant decrease with increasing elevation, whereas endemic richness had two peaks including a major peak around 500 m and a minor peak near 1800 m. Species density had two peaks at mid-elevation zones for endemic and non-endemic fishes, such as 1500–2000 m and 3500–4000 m for endemic fishes and 500–1000 m and 3500–4000 m for non-endemic fishes, respectively. In addition, the structure of endemic fish assemblages in the upper Yangtze River was highly correlated with local topographic and geomorphic characteristics.

The combined effects of pollution, habitat degradation, and overexploitation have reduced fish stocks dramatically. Hydrological alterations are perhaps the largest threat to fish biodiversity in the Yangtze River basin, such as dam construction and disconnection between the river and its lakes. Conservation measures which were applied currently to protect the fish biodiversity in the basin exert ineffectively, such as nature reserve establishment, artificial propagation, and releasing. Therefore, in order to preserve fish biodiversity more effectively in this area, reserve networks, rather than a single national nature reserve, should be

---

L. Huang
College of Environmental Science and Engineering, Guilin University of Technology, Guilin 541004, China

J. Li (✉)
College of Environmental Science and Engineering, Tongji University, Shanghai 200092, China
e-mail: leejianhua@tongji.edu.cn

S.-i. Nakano et al. (eds.), *Aquatic Biodiversity Conservation and Ecosystem Services*, Ecological Research Monographs, DOI 10.1007/978-981-10-0780-4_2

established. Moreover, improved artificial releasing and habitat rehabilitation should be considered urgently.

Keywords Fish diversity • Threats • Conservation • Yangtze River

Introduction

Freshwater fishes face a global crisis, as is evident from the fact that almost 40 % of fishes in Europe and the USA are imperiled (Kottelat and Freyhof 2007; Jelks et al. 2008), which illustrates that freshwater ecosystems tend to have a higher portion of species threatened with extinction than marine or terrestrial counterparts (Revenga et al. 2005; Dudgeon et al. 2006; Strayer and Dudgeon 2010). The Yangtze River basin is a globally significant area for preserving fish diversity and fishery resources due to high fish species richness and endemism in the Palearctic region and its fishery yield accounting for about two-thirds of freshwater fishery production of the whole country (Fu et al. 2003; Liu and Cao 1992; Chen et al. 2002a, b). However, the Yangtze River basin is also an area highly impacted by a long history of human use and environmental variation and is further threatened by the rapid economic development in China and the demands of over 400 million people living in the basin during the past few decades. Since the 1950s, loss of fish diversity in the Yangtze River basin has been accelerated by a series of direct and indirect effects of human activities and environmental changes, such as disappearance, shrinkage and fragmentation of habitats for fish spawning, feeding and migration, overfishing, water pollution, and invasion of exotic species (Fu et al. 2003; Dudgeon et al. 2006). Documenting regional fish distribution and understanding major threats to fish biodiversity are necessary for protecting and recovering endangered fish species and natural communities. Thus, it is urgently needed to review the problems and threats facing fish resources in the Yangtze River basin and to provide crucial information on which species are at risk and what factors threaten their existence for developing successful conservation strategies to slow the loss of fish diversity.

Since the 1930s, Chinese ichthyologists have investigated fish fauna and diversity in the Yangtze River basin, and a large amount of information has been published on taxonomic, biological, and biogeographical aspects. In the present study, we collected and synthesized these scattered data from relevant literature including Anonymous (1976), Zeng (1990), Chen (1998), Yue and Chen (1998), Chu et al. (1999), Yue (2000), Fu et al. (2003), Zhang and Li (2007), Wu and Zhong (2008), Huang et al. (2013), and Li et al. (2013). The threatened status of fish species in the Yangtze River basin was compiled from China Species Red List (Wang and Xie 2009) and China Red Data Book of Endangered Animals (Yue and Chen 1998) into five threatened categories: extinct (EX), extinct in the wild (EW), critically endangered (CR), endangered (EN), and vulnerable (VU). Our main purposes were (1) to investigate large-scale distribution and endemic species composition of fishes in different reaches of the Yangtze River basin, (2) to rank the major threats to impaired or extinct fish species and quantify the relative contribution of intrinsic factors to fish endangerment, and (3) to provide recommendations for fish biodiversity conservation in the Yangtze River basin.

Fish Distribution and Composition

Description of Yangtze River Basin

The Yangtze River basin covers an area of about 1.8 million km^2 and lies across the three large topographic platforms of the Chinese mainland (Fig. 2.1). The riverhead is located on the Tibetan Plateau, where the mean elevation is over 4500 m (Zeng 1990). The geographical range of the Yangtze River basin was defined by Zhu et al. (1998). The basin was further divided into 12 sub-basins based on water systems (Changjiang Hydrological Committee of Hydrology Ministry 1999) (see Fig. 2.1). Upstream from Yichang (in Hubei Province) is the upper reach, with a narrow valley and a rocky channel of high gradient ratio. The length is more than 4300 km, and the drainage area is almost 100×10^4 km^2. The famous Three Gorges Dam is located in the upper mainstream. The range of the middle reach is from Yichang to Hukou (in Jiangxi Province), with a length of about 950 km and a drainage area of almost 68×10^4 km^2. Here, the river has a gentle gradient and takes a meandering course. It connects with shallow lakes of various sizes and numerous tributaries and forms an endemic Chinese compound ecosystem of inland water. From Hukou downward to the mouth of the river is the lower reach, with a length of about 930 km. The drainage area is about 12×10^4 km^2. This segment of the river wanders among plains and hills, and several large interior lakes, such as Lake Chao Hu and Lake Taihu in association with many tributaries, drain into the reach (Fu et al. 2003). In the estuary confluent with the East China Sea, the river forms a trumpet shaped delta.

Faunal Composition

The Chinese freshwater fish fauna numbers well over 1000 species, and at least 717 species in 33 families inhabit rivers (Li 1981); a further 66 species spend part of their lives in rivers, while others are mainly confined to estuarine reaches but occasionally swim upstream. Our synthesis found over 370 species (subordinates to 178 genera, 52 families, 17 orders) known from this basin (including anadromous and catadromous species), of which cyprinids (Cyprinidae) account for 51 %, loaches (Cobitidae) 6.9 %, bagrids (Bagridae) 6.9 %, Homalopteridae 5.5 %, gobies

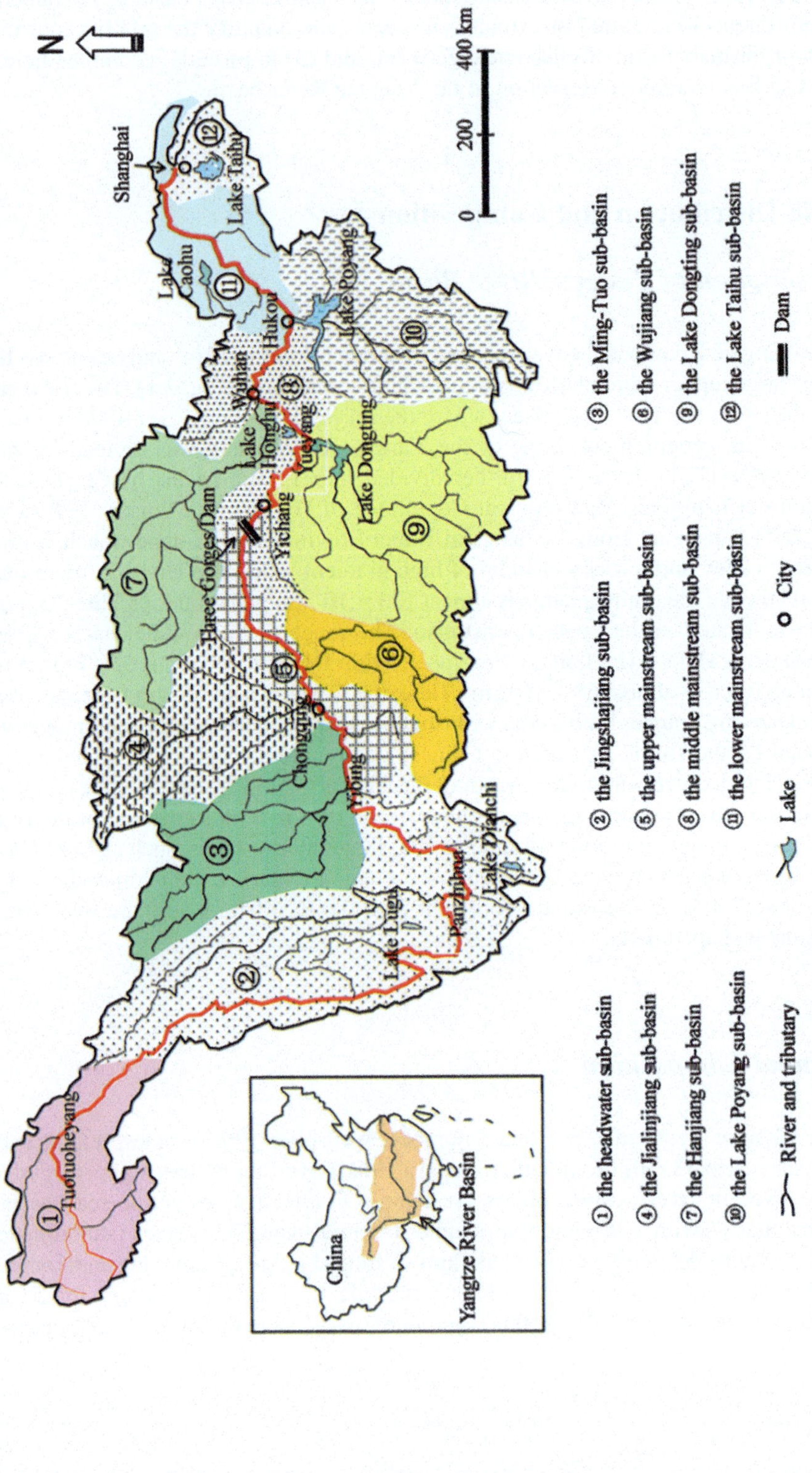

Fig. 2.1 Sketch map of the Yangtze River basin (Cite from Fu et al. 2003)

(Cobitidae) 4.4 %, and other families 24.9 %. Among these are 177 species and subspecies endemic to the basin.

The main species of economic fishes are as follows:

Black carp, *Mylopharyngodon piceus*
Grass carp, *Ctenopharyngodon idellus*
Silver carp, *Hypophthalmichthys molitrix*
Bighead carp, *Aristichthys nobilis*
Common carp, *Cyprinus carpio*
Crucian carp, *Carassius auratus*
Blunt snout bream, *Megalobrama amblycephala*
Whitefish, *Hemisalanx brachyrostralis*
Silurid catfish, *Silurus asotus*
Copper fish, *Coreius heterodon*
Chinese shad, *M. reevesii*
Anchovy, *C. mystus*
Chinese sturgeon, *A. sinensis*
Paddle fish, *Psephurus gladius*
Chinese sucker, *M. asiaticus*
Eel, *Anguilla japonica*

These fish species vary in distribution from section to section of the river (Xie and Chen 1999).

Fish Distribution

Recent evidences suggested that non-endemic richness showed a significant decrease with increasing elevation, whereas endemic richness had two peaks including a major peak around 500 m and a minor peak near 1800 m. Species density had two peaks at mid-elevation zones for endemic and non-endemic fishes, such as 1500–2000 m and 3500–4000 m for endemic fishes and 500–1000 m and 3500–4000 m for non-endemic fishes, respectively (Fu et al. 2004). The Yangtze River basin is usually divided into three parts, i.e., the upper reaches, the middle reaches, and the lower reaches. But the fishes are not evenly distributed in the Yangtze River basin. Common species that occur in two or three reaches, which are less than 25 % of the total fish species number in the basin. Endemic fishes appear to have the similar distribution pattern. Only four endemic species (*Squalidus nitens*, *Saurogobio gymnocheilus*, *Parabotia bimaculata*, and *Pelteobagrus eupogon*) are common to the three reaches.

In the river head waters, the number of fishes is very small, including five cyprinid fishes, seven species of genus *Triplophysa*, and two species of family Sisoridae. Here, the mean elevation is over 4500 m (Zeng 1990) and only some specialists could live. Of these fishes, five species were found only in the riverhead, and the other nine species were also reported in the upper reach. Nearly 60 %

species are endemic fishes, and one species (*Triplophysa tanggulaensis*) was recorded only in the riverhead.

There are 279 species and subspecies of fishes from a wide range of taxonomic categories in the upper reaches. In particular, 124 endemic species were found only in the upper reaches (He et al. 2011), indicating that the distribution of the endemic fishes intensively occurs there. Taking food of attached algae and invertebrates, most of them adapt to rapid current and inhabit the under layer, while others (fishes of families Homalopteridae and Sisoridae) adhere to and climb on the stones of the riverbed. The Chinese paddlefish (*P. gladius*) and Chinese sucker (*M. asiaticus*), which are distributed mainly in the Yangtze River and could be found in all of the three reaches and their main branches, at present occur chiefly in the upper reach. There were only two migratory fishes, but the Japanese eel (*A. japonica*) disappeared after the Gezhouba Dam was constructed in the upper mainstream.

Fishes in the middle and lower reaches are also abundant but with a lower proportion of endemic fishes compared with the upper reaches. More migratory fishes (*A. marmorata*, *M. reevesi*, *C. ectenes*, *C. mystus*, *M. asiaticus*, *Hemisalanx prognathous*, *Trachidermus fasciatus*, *T. obscurus*, *T. flavidus*) occur in the middle and lower reaches. Some euryhaline marine fishes inhabit the lower reaches and the estuary of the river, such as *Neosalanx andersoni*, *Liza carinatus*, *Mugil cephalus*, *Terapon jarbua*, *Repomucenus olidus*, and *Cynoglossus gracilis* (Chen et al. 2004).

The Threats to Fish Diversity

Rivers serve as the primary source of renewable water supply for humans and freshwater ecosystems, yet freshwater systems are considered to be in a state of crisis (Vörösmarty et al. 2010). Estimates suggest that the Earth is in the midst of the sixth mass extinction in its history, with at least 10,000–20,000 freshwater species are to be extinct or at risk (Dirzo and Raven 2003; Strayer and Dudgeon 2010). In the Yangtze River basin, there are 65 threatened fish species included in the China Species Red List (Wang and Xie 2009), belonging to 10 orders and 18 families (Fu et al. 2003). These fishes are classified into 5 threatened categories, i.e., extinct (2 species), extinct in the wild (2 species), critically endangered (5 species), endangered (36 species), and vulnerable (20 species). It should be noted that 69 % of the threatened or extinct fishes are endemic species in the Yangtze River basin. The principal drivers of extinction are habitat loss, overexploitation, pollution, climate change, and impacts of invasive species, which have not only dramatically increased the rate of extinction but have also generally decreased the biodiversity at local scales and the species composition of communities (Dudgeon et al. 2006; Duffy et al. 2009; Secretariat of the Convention on Biological Diversity 2010).

Hydrological Alteration

Hydrological alterations are of great concern to the threat of fish biodiversity in Chinese rivers (Fu et al. 2003; Dudgeon 2011). The main hydrological alterations are building dams on the river, separating lakes from the river, and diverting water.

Dam Construction

The *A. sinensis* with landings of up to 25 t each year and a smaller but significant yield of *A. dabryanus* (Wei 2009); these ceased to be viable fisheries over 30 years ago (Yue and Chen 1998). The *M. asiaticus* formerly contributed more than 10 % of the catch in sections of the Yangtze above the TGD but have dwindled to virtually nothing, and this species vanished from the lower course following construction of the Gezhouba Dam on the Yangtze mainstream during the early 1980s (Yue and Chen 1998; Chen et al. 2004).

Since the foundation of the People's Republic of China, over 50,000 reservoirs of various kinds (including the Gezhouba Hydroelectric project and Three Gorges Dam) have been constructed. Effects of the Gezhouba dam on fish diversity are exemplified by the Acipenseriformes. There are two sturgeons (Chinese sturgeon *A. sinensis* and Dabry's sturgeon *A. dabryanus*) and one paddlefish (Chinese paddlefish *P. gladius*) in the Yangtze River. Spawning migrations of the anadromous Chinese sturgeon were blocked by the Gezhouba Dam. The abundance of the species has greatly declined, despite the re-established spawning habitat below the Gezhouba Dam (Wei et al. 1997). Recent evidence suggests that *P. gladius* are on the verge of extinction or possibly extinct (Wei 2009; Dudgeon 2011): a 3 year hydroacoustic and fisheries survey of the upper Yangtze failed to detect any *P. gladius* despite including the only known recent spawning site of this species (Xie et al. 2007; Zhang et al. 2009). Adults have not been captured since 2003 and there does not appear to be any recruitment of juveniles (Zhang et al. 2009). The situation of *A. dabryanus* seems little better: they are now extremely rare and seldom captured, and only a few small hybrid *A. dabryanus* were encountered during the *P. gladius* survey (Zhang et al. 2009). This point has added significance given that, since 2007, more than 5000 individuals of artificially propagated juveniles have been released into the upper Yangtze River, and this restocking may be all that is maintaining of *A. dabryanus* in the wild (Wei 2009).

Apart from the Gezhouba Dam and the Three Gorges Dam, the government has begun to or is planning to construct at least ten more dams in the upper reaches of the Yangtze River (Dudgeon 2011). The Jinsha 12 dam cascade not only illustrates the lack of commitment to environmental regulations by national and provincial officials, but it also provides a dramatic example of the threats posed to river fishes by China's unfettered economic development (Dudgeon 2011). For example, the China Three Gorges Corporation (CTGC) is building one of its dams, the 161 m tall, 6.4 GW Xiangjiaba Dam, within the former boundaries of the upper Yangtze

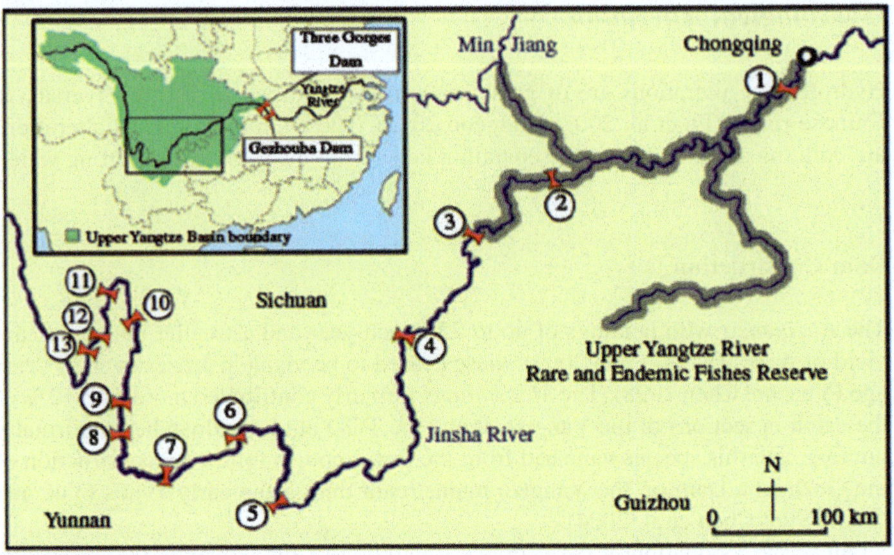

Fig. 2.2 The Jinsha (*upper Yangtze*) River in China showing the location of 13 dams planned or under construction above the Three Gorges Dam (TGD) and Gezhouba Dam (see *inset*). The original boundaries of the Upper Yangtze Rare and Endemic Fishes Reserve, adjacent to Chongqing municipality, are shown ( ). The reserve contains the only known breeding site of *P. gladius* (Dudgeon 2011). *1* Xiaonanhai, *2* Xiangjiaba, *3* Xiluodu, *4* Baihetan, *5* Wudongde, *6* Guanyinyan, *7* Ludila, *8* Longkaikou, *9* Jin'anqiao, *10* Ahia, *11* Liyuan, *12* Liangjiaren, and *13* Hutiaoxia (Cite from Dudgeon 2011)

River Rare and Endemic Fishes Reserve (Fig. 2.2). The 400 km long reserve was designated by the State Council of China in 1987 to protect *P. gladius*, *A. dabryanus*, and 69 endemic or valuable fish species, 29 of which are listed as endangered in the *China Red Data Book* (Wang and Dou 1998), and to offset potential adverse effects of the TGD. The construction of the Wanan Dam in the upper Gan Jiang River in 1986 destroyed the ecological conditions for spawning of the anadromous Chinese shad and almost led to its disappearance in the River (Huang et al. 2013).

River-Lake Separation

Historically, the lakes interlaced with the mainstream and tributaries of the Yangtze River to form a complex lacustrine-riverine network. Fishes could move freely in this system, from one lake to another, through or across the large river, and among tributaries. During the late 1950s–1970s, sluice gates were constructed in almost all lakes, except for Lake Dongting and Lake Poyang for water conservancy projects (Lu et al. 2006; Zhu et al. 2007). The sluice gates blocked interchanges of fishes in the river-lake ecosystem. The fish fauna was formed under the conditions of rivers on the plain as the base. Its development is closely related to the periodic flood and many river-communication lakes distributed on it. Firstly, most fish species have the habit of spawning in running water, laying drifting eggs. The hatched fries enter lakes to live and nourish as they drift with the current. The adults recede from the lakes to the river for wintering or spawning. Secondly, there are many more organisms (plankton, aquatic higher plants, and benthic animals) for fishes in still-water lakes with sludge deposit than in the river (Gong et al. 2011). Thus, lakes are environments especially suitable for nourishment and growth of fry. Most of the fish species, having pelagic coloration of dark back and white belly, are suitable for living in open waters. At the formation of the plain complex, there emerged fish species that had already adapted themselves to biotic and abiotic surroundings in the geographic landscape zone and could propagate themselves generation after generation.

In over 30 years, more than 7000 drainage sluices were constructed in the Yangtze valley (Dudgeon 2011). With the exception of the Dongting Lake and the Poyang Lake, which are still connected with the Yangtze River, all the lakes are isolated from the river. Because of the blockage of pages by the sluices, fish can't migrate from lakes into the river, which resulted in decrease of fish species in those disconnected rivers (Fig. 2.3), such as Taihu Lake and Chao Hu Lake in the lower reaches of Yangtze River basin (Wang 1987; Liao et al. 2002; Fu et al. 2003; Zhang and Li 2007; Guo et al. 2007; Zhu et al. 2007). The fundamental change in ecological environments prevents fish fry and eel elvers from entering lakes for nourishment and growth or spawners from returning to the river or the sea for spawning. Therefore many fish species have been depleted for lack of normal recruit populations. Furthermore, recent evidences suggested that river-lake disconnection reduced fish diversity of Yangtze lakes by 38.1 %, so that the river-connected lakes play an important role in maintaining the floodplain biodiversity (Liu and Wang 2010).

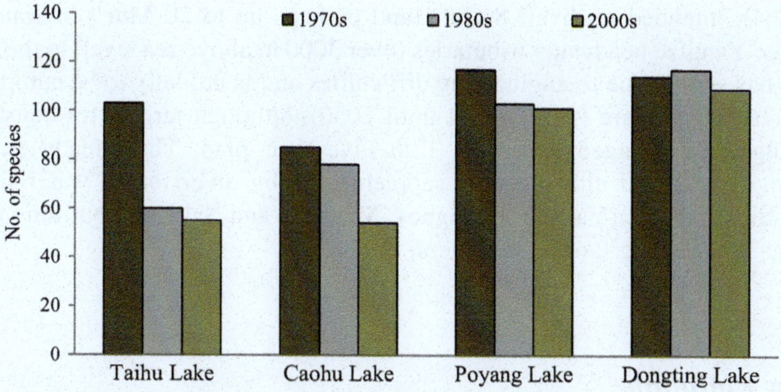

Fig. 2.3 Number of fish species in the disconnected lakes (Taihu Lake and Chao Hu Lake) and connected lakes (Dongting Lake and Poyang Lake) in the Yangtze River basin

Land Reclamation

Lakes are good sites for propagation, habitation, feeding, and growth of fishes. In the past 40 years, under the guidance of the thought of "taking grain as the key line," large areas of land were reclaimed from lakes by building dykes. It was reported that 0.08 million hectares of land were reclaimed from the Poyang Lake (Qian et al. 2002) (a decrease in spawning grounds by 50.0 %) and 0.16 million hectares of land from the Dongting Lake (Liao et al. 2002) (a decrease by 35.8 %). Land reclamation not only destroyed the spawning ground of some resident fish species but also affected the size of recruit population and ultimately resulted in the decline of catch (Fu et al. 2003; Huang et al. 2013).

Water Diversion

With northern China quickly developing, the water deficiency problem in the area is urgent. The government is promoting a plan to divert water to the north area from the Yangtze River basin. It would be the biggest water diversion project in the world. The project has been stalled for decades, but initiating construction was a key component of China's 10th Five-Year plan (2001–2005). Work began in 2002 on an eastern route (Fig. 2.4) along the coast through the ancient, but now refurbished, Grand Canal that will transfer (polluted) water from the lower Yangtze. Upon completion in 2013, 14.8 Gm^3 of water will be conveyed as far as Tianjin, almost 1200 km north, each year. A central middle route (Fig. 2.4) transferring water from the Danjiangkou Reservoir on a major Yangtze tributary (the Han River) has been under construction since 2003 and, by 2014, should be transferring 13 Gm^3 of water annually 1300 km north to Beijing, Tianjin, and other northern cities. An additional southern connection to the TGD has been suggested, so as to reduce the amount of water removed from the Han River. A western route (Fig. 2.4), intended to divert 8 Gm^3 (and perhaps up to 20 Mm^3) of water from the three Yangtze headwater tributaries (over 3000 m above sea level) to the Yellow River, has stalled due to engineering difficulties and is unlikely to be completed in the foreseeable future (perhaps not until 2050), although targets for construction planning are envisaged under the 12th Five-Year plan. The already ambitious western route could ultimately be supplemented by diversion of water from the upper Brahmaputra (Yarlung Zhangbo), Mekong, and Salween, but thus far, this remains engineering fantasy (Dudgeon 2011).

Water Pollution

Over 400 million people live in the 1.8 million km^2 Yangtze basin, and it is the source of 40 % of the country's GDP (Pittock and Xu 2010). For that reason, the Yangtze receives around half of China's wastewater discharge (Dudgeon 2010),

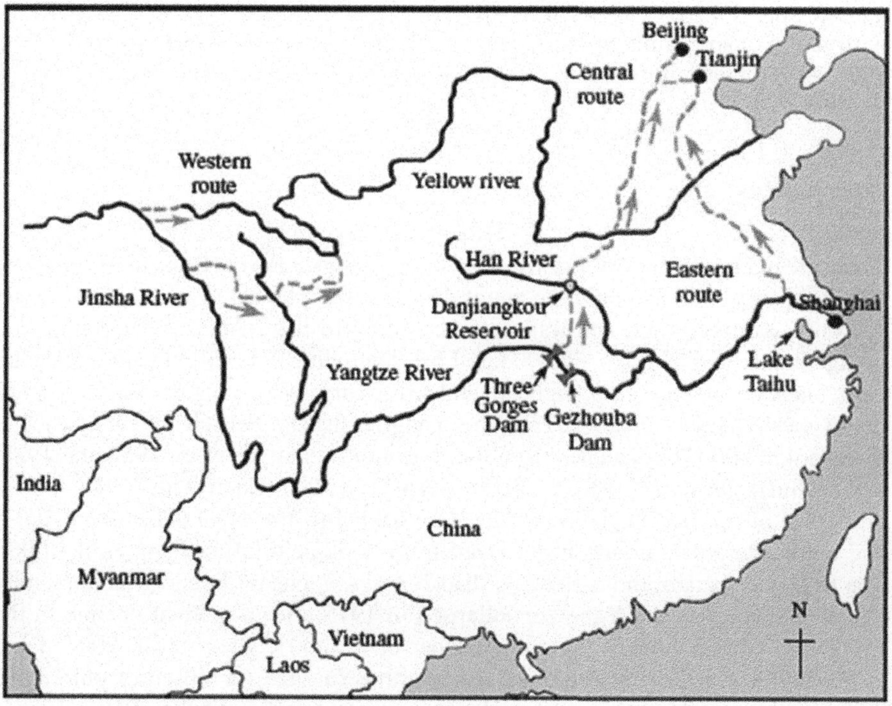

Fig. 2.4 The eastern, western, and central routes intended to transfer water from the Yangtze River to the north of China (Cite from Dudgeon 2011)

and point-source pollution by sewage and industrial waste is compounded by nitrogen runoff (almost four times the global average) as well as phosphorus and pesticides from agricultural land and contaminants from vessels (Li and Zhang 1999; Xue et al. 2008). Ambient levels of persistent organic pollutants and polycyclic aromatic hydrocarbons from a variety of sources have increased along the river (Xue et al. 2008), and sublethal effects such as impairment of reproduction have been documented for *A. sinensis* (Hu et al. 2009a, b). The Chinese Ministry of Environmental Protection (MEP) annual reports (MEP 2008) show that pollution burdens in parts of the Yangtze have increased in recent years, especially in the lower course and in smaller tributaries (Xue et al. 2008). Only 31 % of water samples, mainly from the upper Yangtze, are of class I or II quality, which can be for drinking water sources, and much of the lower Yangtze water is Class III or poorer, which cannot be for drinking water sources. There is, however, evidence of improvements in some localities due to enhanced treatment of industrial wastewater (MEP 2008). Therefore, the contamination of water in the river has become more and more severs. On the whole, the water quality is quite good. But in certain sections, water is gravely contaminated. According to incomplete data, in 21 cities along the river, there are over 2000 spot sources, discharging 14.2 billion tons of waste in 2000 (Chen et al. 2002b; Guan et al. 2011). The discharge of wastewater and sewage into the river resulted in the pollution of waters and food organisms, destruction of spawning grounds, depletion of brood stocks, decrease in production, and even high fish mortality in certain sections of the Yangtze River (Fu et al. 2003).

Overfishing

There are over 160 kinds of fishing gears in the Yangtze River (Duan et al. 2002), of which the most harmful are fyke net of dense mesh, fyke net of ramie cloth, maze, drop net, damming net, and trap net. According to the statistical data, in Banhu village in the Poyang Lake, 85,000 kg of fish (including 0.3 million young common carp, black carp, grass carp, bighead, and silver carp) were caught with fyke net of dense mesh (Huang et al. 2013). The Yangtze fishery peaked in 1954 when it yielded 450,000 t, but catches from the river fell by half between 1950 and 1970, and declined subsequently to *c*. 130,000 t $year^{-1}$ in 2000 (Chen et al. 2004). Almost all (97 %) of this yield came from the Yangtze below the TGD (Chen et al. 2004) and a considerable amount was derived from floodplain lakes (Fang et al. 2006). The fishery for anadromous Reeve's shad *T. reevesii*, one of the most economically valuable species in the Yangtze, collapsed in 1975 after a gradual decline in the mean size of individuals captured (Blaber et al. 2003; Chen et al. 2004; Fang et al. 2006). Another anadromous species, the Yangtze (or obscure) puffer fish *T. fasciatus*, sustained a commercially important fishery until the 1960s (Turvey et al. 2010). In addition to overexploitation, particularly by fine-meshed nets, the decline of *T. reevesii* can be attributed to dams on Yangtze tributaries, such that this fish is classified as endangered in the *China Red Data Book* with fishing of juveniles prohibited (but poorly enforced; Yue and Chen 1998). Declines in the long-snouted catfish *Leiocassis longirostris* are probably due to overfishing as the dried swim bladder is highly prized; landings have diminished to such an extent that it has almost disappeared from markets (Chen et al. 2004).

Inland Navigation

In large rivers or lakes under regular use as shipping routes, the issue of how disturbance induced by navigation affects fish assemblage has been often discussed. Wolter and Arlinghaus (2003) suggested a navigation bottleneck hypothesis (NBH) in which they mainly focused on chronic physical water movement which exceeds the fish's swimming ability to counteract (Killgore et al. 2001; Gutreuter et al. 2003). Navigation also increases water turbidity (Kucera-Hirzinger et al. 2009), where chronic turbid water affects growth and survival rate (Sutherland and Meyer 2007; Rosso et al. 2010) as well as other behaviors such as feeding (Kano et al. 2011) of fish individuals. Macrophytes are also impacted by navigation activities, which, in turn, affect their ecological functions as fish habitats and shelters (Willby and Eaton 1996). The impact upon survival of eggs or young is because navigation disturbances prevent fish from nest-guarding (Mueller 1980) or can dislodge eggs (Jude et al. 1998). Therefore, navigation in the Yangtze River would not only affect the fish diversity and other aquatic animals directly (Dong et al. 2012) but also pollute the water where the ship went and affect fish diversity indirectly (Pu 2010).

Conservation Strategies

The current conservation policies for freshwater fish biodiversity in China mainly focus on the endangered fishes and economic fishes, while ordinary fish are not of great consideration. Since the Yangtze River basin is undergoing a very rapid deterioration as a consequence of human-induced changes, conservation strategies must be implemented and expanded.

Nature Reserve Establishment

Nature reserve establishment is a good way to conserve where a large number of endemic species were contained in a relatively small area (Myers et al. 2000). A lot of nature reserves have been established in the Yangtze River Basin, and 11 nature reserves have been listed in Ramsar Convention. Current research suggested that structure of the endemic fish assemblages in the upper Yangtze River was highly correlated with local topographic and geomorphic characteristics. Simultaneously, the catchment land cover features also reflected this endemic fish distribution structure (He et al. 2011). Therefore, in order to preserve the rare and endemic fish in the upper Yangtze River, reserve networks, rather than a single national nature reserve, should be established. At present, there are three fish reserves, of which one is for the protection of rare and unique species in the upper reach at the national level and two are for the protection of Chinese sturgeon at the provincial level. The National Nature Reserve for rare and peculiar fish species in the upper Yangtze River is the only national protected zone across many provinces in China. Its management and protection tasks are very arduous. The effective realization of nature reserve functions can be pushed forward only through their tracing, monitoring and appraisal.

Strengthening Administration of Yangtze Fishery

The Yangtze fishery production is very unfavorable to the development of fishery production. Fishing gears and methods that damage juvenile fish stocks should be banned, mesh sizes for various kinds of nets strictly restricted, and the allowable catching sizes of various fish species set so that the fish resources can be replenished. In order to enhance the superiority of the Yangtze River basin in fisheries, it is necessary to establish and strengthen fishery administration agencies at various levels in the Yangtze River basin, which can make and enforce all kinds of laws and regulations on fisheries, implement and inspect plans in fishery production, maintain fishery production order in rivers and lakes, protect fishery resources, and finally intensify the scientific management of fishery resources (Xu et al. 1999).

Implementation of Closed Areas and Seasons to Intensify Propagation Protection

Under the present conditions, the brood and juvenile stocks of shad should be taken as the main protected objects and the protection of the spawning grounds earnestly intensified so as to ensure that the juvenile stocks will not be destroyed (Hu et al. 2009b). The capture of eel should be kept under control and their resources protected. Closed fishing seasons were legislated in different areas of Yangtze River basin (Table 2.1), which pose positive effects on fish resources recovery (Fu et al. 2003; Huang et al. 2013; Li et al. 2013). Closed seasons should be implemented in order to intensify the propagation protection of all fishes in the whole river in the spring.

Rehabilitation of Habitat for Fish

The aquatic ecosystem consists of many sophisticated components that are relevant to and interactive with each other. The variation in each component will influence the overall ecosystem and finally result in the dynamics of fish resources. Over 30 years of practice gave us a profound lesson that the isolation of lakes from rivers, the damming of rivers, land reclamation from the lakes, water pollution, and overfishing destroyed the ecological conditions for fishes and ultimately resulted in a severe depletion of fish stocks in the Yangtze River (Fu et al. 2003; Dudgeon

Table 2.1 Closed fishing season in different areas of Yangtze River basin

Location	Period
Upper reaches of Yangtze River	February 1–April 30
Middle-lower reaches of Yangtze River	April 1–June 30
Dongting Lake	March 10–June 30
Poyang Lake	March 20–June 20
Honghu Lake	April 1–June 30
Chao Hu Lake	February 1–July 31
Taihu Lake	February 1–August 31

2011). Hence, in order to develop fishery production and benefit the people in the Yangtze valley, emphasis should be laid on the comprehensive renovation of waters. Hydraulic architectures should be constructed in protection of the ecological environments for fishes. Wastewater should be discharged in accordance with the criteria formulated by the state. Land reclamation should be forbidden from lakes and detailed regulations on management should be made in order to protect fish resources. For example, the minimum protected area of river-connected lakes was estimated to be 14,400 km^2 in the floodplain of Yangtze River (Liu and Wang 2010). In addition, water pollution in the Yangtze River basin should be paid more attention in the next decades.

Conservation of freshwater biodiversity is perhaps the ultimate conservation challenge because it is influenced by the upstream drainage network, the surrounding land, the riparian zone, and – in the case of migrating aquatic fauna – downstream reaches. Such prerequisites are hardly ever met. Immediate action is needed where opportunities exist to set aside intact lake and river ecosystems within large protected areas. For most of the global land surface, trade-offs between conservation of freshwater biodiversity and human use of ecosystem goods and services are necessary.

References

Anonymous, Department of Ichthyology, Institute of Hydrobiology, CAS (1976) Fishes of the Yangtze River. Science Press, Beijing (in Chinese)

Blaber SJ, Milton DA, Brewer DT, Salini JP (2003) Biology, fisheries, and status of tropical shads *Tenualosa* spp. in South and Southeast Asia. Am Fish Soc Symp 35:49–58

Changjiang Hydrological Committee of Hydrology Ministry (1999) Atlas of the Yangtze River Basin. China Map Press, Beijing, China (in Chinese)

Chen Y (1998) Fauna Sinica: Osteichthyes Cypriniformes II. Science Press, Beijing (in Chinese)

Chen DQ, Liu SP, Duan XB (2002a) A preliminary study of the fisheries biology of main commercial fishes in the middle and upper reaches of the Yangtze River. Acta Hydrobiol Sin 26:622–626 (in Chinese)

Chen JC, Sun ZZ, Qu JH (2002b) Water pollution and its toxicity to fishes in the major areas of the lower reaches of the Yangtze River. Acta Hydrobiol Sin 26:640–646 (in Chinese)

Chen D, Duan X, Liu S, Shi W (2004) Status and management of the fisheries resources of the Yangtze River. In: Welcomme R, Petr T (eds) Proceedings of the second international symposium on the management of large rivers for fisheries, vol 1. FAO Regional Office for Asia and the Pacific, Bangkok, pp 173–182

Chu X, Zheng B, Dai D (1999) Fauna Sinica: Osteichthyes. Siluriformes. Science Press, Beijing (in Chinese)

Dirzo R, Raven PH (2003) Global state of biodiversity and loss. Ann Rev Environ Resour 28:137–167

Dong SY, Dong LJ, Li SH, Kimura S, Akamatsu T, Wang KX, Wang D (2012) Effect of vessel traffic on the acoustic behavior of Yangtze finless porpoises (*Neophocaena phocaenoides asiaeorientalis*) in the confluence of Poyang Lake and the Yangtze River: using fixed passive acoustic observation methods. Acta Hydrobiol Sin 36:246–254 (in Chinese)

Duan XB, Chen DQ, Liu SP (2002) Studies on status of fishery resources in Three Gorges Reservoir reaches of the Yangtze River. Acta Hydrobiol Sin 26:611–616 (in Chinese)

Dudgeon D (2010) Requiem for a river: extinctions, climate change and the last of the Yangtze. Aquat Conserv-Mar Freshw Ecosyst 20(2):127–131

Dudgeon D (2011) Asian river fishes in the Anthropocene: threats and conservation challenges in an era of rapid environmental change. J Fish Biol 79:1487–1524

Dudgeon D, Arthington AH, Gessner MO, Kawabata Z, Knowler D, Lévêque C, Naiman RJ, Prieur-Richard AH, Soto D, Stiassny MLJ, Sullivan CA (2006) Freshwater biodiversity: importance, threats, status and conservation challenges. Biol Rev 81:163–182

Duffy JE, Srivastava DS, McLaren J, Sankaran M, Solan M, Griffin J, Emmerson M, Jones KE (2009) Forecasting decline in ecosystem services under realistic scenarios of extinction. In: Naeem S, Bunker DE, Hector A, Loreau M, Perrings C (eds) Biodiversity, ecosystem functioning and human wellbeing: an ecological and economic perspective. Oxford University Press, Oxford, pp 60–77

Fang J, Wang X, Zhao S, Li Y, Tang Z, Yu D, Ni L, Liu H, Xie P, Da L, Li Z, Zheng C (2006) Biodiversity changes in the lakes of the central Yangtze. Front Ecol Environ 4:369–377

Fu CZ, Wu JH, Chen JK, Wu QH, Lei GC (2003) Freshwater fish biodiversity in the Yangtze River basin of China: patterns, threats and conservation. Biodivers Conserv 12:1649–1685

Fu CZ, Wu JH, Wang XY, Lei GC, Chen JK (2004) Patterns of diversity, altitudinal range and body size among freshwater fishes in the Yangtze River basin, China. Glob Ecol Biogeogr 13:543–552

Gong GC, Liu KK, Chiang KP, Hsiung TM, Chang J, Chen CC, Hung CC, Chou WC, Chung CC, Chen HY, Shiah FK, Tsai AY, Hsieh CH, Shiao JC, Tseng CM, Hsu SC, Lee HJ, Lee MA, Lin II, Tsai FJ (2011) Yangtze River floods enhance coastal ocean phytoplankton biomass and potential fish production. Geophys Res Lett 38:L13603. doi:10.1029/2011GL047519

Guan BH, An SQ, Gu BH (2011) Assessment of ecosystem health during the past 40 years for Lake Taihu in the Yangtze River Delta, China. Limnology 12:47–53

Guo L, Xie P, Ni L, Hu W, Li H (2007) The status of fishery resources of Lake Chaohu and its response to eutrophication. Acta Hydrobiol Sin 31:700–705 (in Chinese)

Gutreuter S, Dettmers JM, Wahl DH (2003) Estimating mortality rates of adult fish from entrainment through the propellers of river tow boats. Trans Am Fish Soc 132:646–661

He Y, Wang J, Lek S, Cao W, Lek-Ang S (2011) Structure of endemic fish assemblages in the upper Yangtze River Basin. River Res Appl 27:59–75

Hu J, Zhang Z, Wei Q, Zhen H, Zhao Y, Peng H, Wan Y, Giesy JP, Li L, Zhang B (2009a) Malformations of the endangered Chinese sturgeon, *Acipenser sinensis*, and its causal agent. Proc Natl Acad Sci U S A 106:9339–9344

Hu ML, Wu ZQ, Liu YL, Liu XT, Deng SS (2009b) Analysis on catching fry of four major Chinese carps in the Ruichang section of Yangtze River. Acta Hydrobiol Sin 33:136–139 (in Chinese)

Huang LL, Wu ZQ, Li JH (2013) Fish fauna, biogeography and conservation of freshwater fish in Poyang Lake Drainage, China. Environ Biol Fish 96:229–1243

Jelks HL, Walsh SJ, Burkhead NM, Contreras-Balderas S, Díaz-Pardo E, Hendrickson DA, Lyons J, Mandrak NE, McCormick F, Nelson JS, Platania SP, Porter BA, Renaud CB, Schmitter-Soto JJ, Taylor EB, Warren ML (2008) Conservation status of imperiled North American freshwater and diadromous fishes. Fisheries 33:372–407

Jude DJ, Tesar FJ, Tin HT (1998) Spring distribution and abundance of larval fishes in the st. Marys River, with a note on potential effects of freighter traffic on survival of eggs and larvae. J Great Lakes Res 24:569–581

Kano Y, Sato T, Huang LL, Wood C, Bessho K, Matsumoto T, Shimatani Y, Nakajima J (2011) Navigation disturbance and its impact on fish assemblage in the East Tiaoxi River, China. Landsc Ecol Eng. doi:10.1007/s11355-011-0181-0

Killgore KJ, Maynord ST, Chan MD, Morgan RP (2001) Evaluation of propeller-induced mortality on early life stages of selected fish species. N Am J Fish Manag 21:947–955

Kottelat M, Freyhof J (2007) Handbook of European freshwater fishes. Kottelat & Freyhof, Cornol

Kucera-Hirzinger V, Schludermann E, Zornig H, Weissenbacher A, Schabuss M, Schiemer F (2009) Potential effects of navigation induced navigation induced wave wash on the early life history stages of riverine fish. Aquat Sci 71:94–102

Li S (1981) Studies on zoogeographical divisions for freshwater fishes of China. Science Press, Beijing (in Chinese)

Li Y, Zhang J (1999) Agricultural diffuse pollution from fertilisers and pesticides in China. Water Sci Technol 39:25–32

Li JH, Huang LL, Sato T, Zou LM, Jiang K, Yahara T, Kano Y (2013) Distribution pattern, threats and conservation of fish biodiversity in the East Tiaoxi, China. Environ Biol Fishes 96:519–533

Liao F, He W, Huang X, Jing Q, He X (2002) Studies on present situation and change trend of Dongting Lake fishery resources and environment. Acta Hydrobiol Sin 26:623–627 (in Chinese)

Liu J, Cao W (1992) Fish resources in the Yangtze basin and the strategy for their conservation. Resour Environ Yangtze Valley 1:17–23 (in Chinese)

Liu XQ, Wang HZ (2010) Estimation of minimum area requirement of river-connected lakes for fish diversity conservation in the Yangtze River floodplain. Divers Distrib 16:932–940

Lu S, Hu J, Xiao C, Hu H (2006) Fish species composition in Honghu Lake and estimation of influencing factors. Chin Wildl 27:14–17 (in Chinese)

MEP (2008) Report on the state of the environment in China 2007. Water environment. Ministry of Environmental Protection, Beijing. Available at http://english.mep.gov.cn/standards_reports/soe/soe2007/200909/t20090922_161251.htm/. Accessed 25 Apr 2011.

Mueller G (1980) Effects of recreational river traffic on nest defence by longear sunfish. Trans Am Fish Soc 109:248–251

Myers N, Mittermeier RA, Mittermeier CG, Da Fonseca GAB, Olivieri S (2000) Biodiversity hotspots for conservation priorities. Nature 404:853–858

Pittock J, Xu M (2010) World resources report case study. Controlling Yangtze River Floods: a new approach. World Resources Report, Washington, DC

Pu HF (2010) The prevention and control of pollution caused by navigation in the Yangtze River and its. Express Water Resour Hydropower Inf 4:7–10

Qian XE, Huan CG, Wang YM (2002) The status of fishery resources of Poyang Lake and its environmental monitoring. Acta Hydrobiol Sin 26:617–621 (in Chinese)

Revenga C, Campbell I, Abell R, de Villiers P, Bryer M (2005) Prospects for monitoring freshwater ecosystems towards the 2010 targets. Phil Trans R Soc B 360:397–413

Rosso JJ, Sosnovsky A, Rennella AM, Quiros R (2010) Relationships between fish species abundances and water transparency in hypertrophic turbid waters of temperate shallow lakes. Int Rev Hydrobiol 95:142–155

Secretariat of the Convention on Biological Diversity (2010) Global biodiversity outlook 3, vol 94. CBD, Montreal

Strayer DL, Dudgeon D (2010) Freshwater biodiversity conservation: recent progress and future challenges. J N Am Benthol Soc 29:344–358

Sutherland AB, Meyer JL (2007) Effects of increased suspended sediment on growth rate and gill condition of two southern Appalachian minnows. Environ Biol Fishes 80:389–403

Turvey ST, Barrett LA, Hao Y, Zhang L, Zhang X, Wang X, Huang Y, Zhou K, Hart T, Wang D (2010) Rapidly shifting baselines in Yangtze fishing communities and local memory of extinct species. Conserv Biol 24:778–787

Vörösmarty CJ, McIntyre PB, Gessner MO, Dudgeon D, Prusevich A, Green P, Glidden S, Bunn SE, Sullivan CA, Liermann CR, Davies PM (2010) Global threats to human water security and river biodiversity. Nature 467:555–561

Wang Q (1987) Study on ichthyological fauna of Chao Lake. J Anhui Univ (Nat Sci) 2:70–78 (in Chinese)

Wang S, Xie Y (2009) China species red list. Vol. II vertebrates – part 1. High Education Press, Beijing (in Chinese)

Wang S, Dou H (1998) Lakes, Sinica. Science Press, Beijing (in Chinese)
Wei Q, Ke F, Zhang J, Zhuang P, Luo J, Zhou R, Yang W (1997) Biology, fisheries and conservation of sturgeons and paddlefish in China. Environ Biol Fish 48:241–255
Wei QW (2009) Acipenser dabryanus. IUCN Red list of threatened species, version 2010.4. Available at www.iucnredlist.org. Accessed 22 Apr 2011
Willby NJ, Eaton JW (1996) Backwater habitats and their role in nature conservation on navigable waterways. Hydrobiologia 340:333–338
Wolter C, Arlinghaus R (2003) Navigation impacts on freshwater fish assemblages: the ecological relevance of swimming performance. Rev Fish Biol Fish 13:63–89
Wu H, Zhong J (2008) Fauna Sinica: Osteichthyes. Perciformes V. Gobioidei. Science Press, Beijing (in Chinese)
Xie P, Chen Y (1999) Threats to biodiversity in Chinese inland waters. Ambio 28:674–681
Xie S, Li Z, Liu J, Xia S, Wang H, Murphy BR (2007) Fisheries of the Yangtze River show immediate impacts of the Three Gorges Dam. Fisheries 32:343–344
Xu H, Wang S, Xue D (1999) Biodiversity conservation in China: legislation, plans, and measures. Biodivers Conserv 8:819–837
Xue XM, Xu R, Zhang BR, Li FT (2008) Current pollution status of Yangtze River and controlling measures. Int J Environ Waste Manag 2:267–278
Yue P (2000) Fauna Sinica: Osteichthyes. Cypriniformes III. Science Press, Beijing (in Chinese)
Yue P, Chen Y (1998) Pisces. In: Wang S (ed) China red data book of endangered animals. Science Press, Beijing (in Chinese)
Zeng X (1990) Fishery resources of the Yangtze River Basin. Marine Press, Beijing (in Chinese)
Zhang T, Li Z (2007) Fish resources and fishery utilization of Lake Poyang. J Lake Sci 19:434–444 (in Chinese)
Zhang H, Wei Q, Du H, Shen L, Li Y, Zhao Y (2009) Is there evidence that the Chinese paddle fish (*Psephurus gladius*) still survives in the upper Yangtze River? Concerns inferred from hydroacoustic and capture surveys, 2006–2008. J Appl Ichthyol 25(Suppl. 2):95–99
Zhu Y, He Y, She Z (1998) Atlas of study on background value of aquatic environment of the Changjiang River Valley. Science Press, Beijing (in Chinese)
Zhu S, Liu W, Gu X (2007) Changes of the fish fauna and fish yield analysis in Lake Taihu. J Lake Sci 19:664–669 (in Chinese)

Chapter 3
Freshwater Fish Diversity in Thailand and the Challenges on Its Prosperity Due To River Damming

Tuantong Jutagate, Chaiwut Grudpan, and Apinun Suvarnaraksha

Abstract Thailand is among the countries that ranked top in freshwater fish diversity, of which 843 freshwater fish species, including 13 elasmobranch fishes, have been officially recorded. Presently, the integrity of freshwater fish community is threatened by many stressors from human activities, in particular the river damming. In this paper, we illustrate the impacts to freshwater fish diversity by various dam types in Thailand. The major serious impact of river damming is the blockage of fish migratory routes, both for the species that migrate within river system and migrate between the river and the sea. So far, there is no any fish passage that reveals the effective performance because it is difficult to deal with the many fish species, with various modes of swimming and water flow preferences, in a single designed fish passage. The consequent results are the extirpation of the migratory species and, then, changes in fish community structures. The mitigation measure of opening the sluice gates for a certain period is recommended. This measure is beneficial for the fish migrations both for the dams in the middle portion and lower portion, i.e., brackish area, of the river, which allows the fish to complete their life cycles and can manipulate the fish assemblage patterns.

Keywords Damming • Freshwater fish • Migration • Fish passage • Mitigation • Thailand

---

T. Jutagate (✉) • C. Grudpan
Faculty of Agriculture, Ubon Ratchathani University, Warin Chamrab, Ubon Ratchathani 34190, Thailand
e-mail: tuantong.j@ubu.ac.th

A. Suvarnaraksha
Faculty of Fisheries Technology and Aquatic Resources, Maejo University, Sansai, Chiangmai 50290, Thailand

S.-i. Nakano et al. (eds.), *Aquatic Biodiversity Conservation and Ecosystem Services*, Ecological Research Monographs, DOI 10.1007/978-981-10-0780-4_3

Introduction

It has been reported that Thailand locates in the Indo-Burma biodiversity hotspot and is ranked top in freshwater fish diversity, which has 843 species, including 13 elasmobranch fishes (Fig. 3.1, Froese and Pauly 2012). The number of indigenous species is as high as 661 species and the percentage of endemic is 2.3 (Nguyen and De Silva 2006). Meanwhile, the numbers of critically endangered, endangered, and vulnerable species of freshwater fishes are 13, 29, and 71, respectively (Office of Natural Resources and Environmental Policy and Planning 2013). Cyprinid fish is the hugest species which accounts for about 44 % of overall freshwater taxon and ecologically diverse group (Froese and Pauly 2012). This exceptional fish diversity, incorporated with an extensive inland water area (i.e., about 4.5×10^4 km^2), within seven major river basins, supports a huge fishery of about 2.2×10^5 tonnes/year^{-1} and contributes to about 7 % of the country's overall fish productions (Jutagate 2009). Suvarnaraksha et al. (2012) demonstrated that the assemblage patterns along the longitudinal gradient of rivers in Thailand can be classified according to their habitat preferences:

(a) The "mountainous" species such as *Oreoglanis siamensis*, *Exostoma vinciguerra*, *Glyptothorax trilineatus*, and *Devario maetaengensis*.
(b) The "piedmont" species such as *Lepidocephalichthys hasselti*, *Dermogenys pusilla*, *Channa gachua*, and *Homaloptera leonardi*.
(c) The "transitory" species (i.e., move between the piedmont and lowland) such as *Puntius orphoides*, *Mastacembelus armatus*, *Puntius brevis*, and *Mystacoleucus marginatus*.
(d) The "lowland" species, which are fishes of family Cyprinidae (many species), Pangasiidae, and Notopteridae.

The integrity of freshwater fish community and sustainability of the inland fisheries are in question, in particular the riverine inhabitants, because of many stressors to rivers, especially river damming. There are a number of damming projects both in terms of "run-of-the-river" and "lake" types, which can be classified according to their purposes and locations as (a) river damming for the irrigation or hydropower purposes in the northeast, (b) building of the "anti-salt" dam in the south, and (c) river damming for flood protection and irrigation in the north and central. Although these infrastructure developments have numerous economic benefits, the pros and cons are on-going, especially in terms of fish and fisheries. As for the pros, fishery is always considered as secondary use of the impounded water regardless of the main purpose of the damming (De Silva and Amarasinghe 2009). For example, in the lower Mekong Basin, fishery yield of large reservoirs generally ranged between 100 and 300 kg ha^{-1} year^{-1} (Hortle and Bamrungrach 2012) and ranged from less than 50 and up to 500 kg ha^{-1} year^{-1} in Thailand (Jutagate 2009).

As for the cons, there are also numbers of literatures describing their impacts on fish populations (e.g., Marmulla 2001; Welcomme et al. 2006), for example, dams

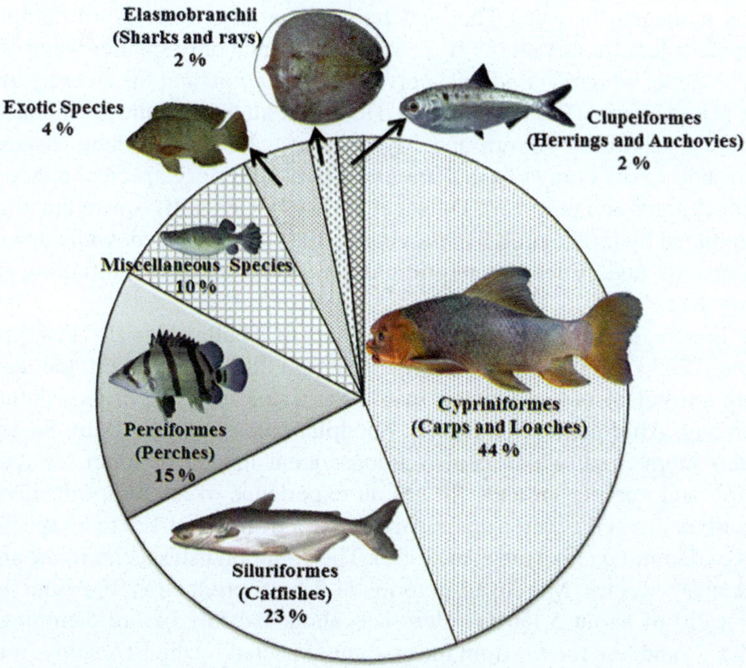

Fig. 3.1 Percentage of species composition for freshwater fishes in Thailand

interrupt the river flow, generate hydrological changes along the integrated continuum of river ecosystems, and form new lentic or semi-lentic environments, which eventually impact to the assemblage patterns of fish inhabiting both upstream and downstream of dams. Moreover, the dam itself can block or delay the migrations of the potamodromous (migrate within the river system) and diadromous (migrate between rivers and the sea) species and thus contribute to the decline and even the extinction of species (Larinier 2001; Welcomme et al. 2006). In this paper, we reviewed the experiences on river damming to fishes and fisheries in Thailand, according to its type, as well as their mitigation measures on the fishery resources. The results would eventually be a lesson learned to look for the challenge to maintaining the integrity of the aquatic ecosystem and enhancing reservoir fisheries.

"Run-of-the-River-Type" Damming to Fish Community

In Thailand, damming the river as "run-of-the-river type" had been built for irrigation (e.g., Chao Phraya Dam in Chainat Province and the cascade dams of Gam River in Sakon Nakhon Province) or hydroelectric generation (e.g., Pak Mun Dam in Ubon Ratchathani Province and the Ban Kum hydropower project on the

Mekong mainstem between Thai and Lao). Generally, this type of damming has considerably less impact on the river ecosystem since only smaller water is stored, or maybe none, which is used to supply a power station, and the river hydrological regime is less altered (Raja et al. 2006). However, all "run-of-the-river" dams act as a barrier to migration (Baran and Myschowoda 2009), and many fishes of the Mekong and Chao Phraya basins are known as migratory species, although with different degrees and purposes (Dugan et al. 2010). Generally, upstream migrations are dominated by larger adult fish moving up the rivers to breed, while downstream migrations are mainly feeding migrations for larvae and adults returning from the breeding areas.

The lesson learned of this damming type, in Thailand, is the Pak Mun dam (Roberts 1995, 2001), which was built across the Mun River, the largest of Mekong's tributary of the country, in 1990 at 6 km upstream of the confluence to the Mekong. After damming, the vast floodplain area along the Mun River's bank had been submerged, which is also a consequent loss in its main service as fish spawning and nursing grounds. It was an experience of an abrupt decline of fish species diversity in the dam's upstream area, i.e., from about 120 to 59 species, after 5 years of damming (Jutagate et al. 2001). The common fish species in the area were the sedentary species. Moreover, in terms of fishery production, the reductions in a fishery yield of about 5 kg ha^{-1} was less about twofold before damming (about 10 kg ha^{-1}) and less too far from the amount expected by the EIA study (more than 50 kg ha^{-1}). There were very few number of the long migratory fish found in the upstream area, although there is the fish ladder attached to the dam (Roberts 2001; Jutagate et al. 2001).

To mitigate this problem, the Royal Thai government has launched a measure that all sluice gates be opened for 4 months (around mid-June to mid-October) and then closed for 8 months for hydropower generation, according to the scientific findings and recommendations after the 1-year trial opening all sluice gates of the Pak Mun Dam (details in Jutagate et al. 2003, 2005). This mitigation measure for fishes and fisheries impacted by hydropower project has never been applied anywhere (Trussart et al. 2002) but the Pak Mun Dam. The period of sluice gate opening coincides with the period of the reversed flow up from the Mekong mainstream to the Mun River that makes hydropower generation unworkable. The followed up studies showed the benefit of this measure for potamodromous fish in terms of spawning by tracing the changes in the maturity stages and the annual sex hormonal profiles between the fishes in the upstream and downstream of dam (details in Jutagate and Krudphan 2004; Jutagate et al. 2007).

Benefits of this mitigation measure for individual fish species (or groups) were also tested by applying the time series data on the presence-absence data of 210 fish species in the upstream of the Pak Mun Dam before, during, and after the dam construction and during the 1-year trial of opening all sluice gates (Jutagate et al. 2008) as well as the data from the monitoring program after the measure has been implemented between 2004 and 2006. The fish assemblage patterns during each period were classified by the occurrence probability of each species in each period (Jutagate et al. 2008), which found that the fishes in the lotic and semi-lotic guilds, i.e., the fish species that are generally longitudinal migrants that move within the main river channel or up and down tributaries, were the most beneficial groups due to this mitigation measure, such as *Cosmochilus harmandi*, *Hypsibarbus* spp., *Mekongina erythrospila*, *Hemisilurus mekongensis*, *Helicophagus leptorhynchus*, and *Pangasius* spp.

"Reservoir-Type" Damming to Fish Community

In general, after impoundment, there are changes in fish communities because of the strong alterations of physical and chemical properties, as well as biological productivity of ecosystems. The long-term dynamic patterns of fish communities in large Thai reservoirs (Jutagate et al. 2012) showed that some riverine origin fishes such as *Clupeichthys aesarnensis*, *Henicorhynchus siamensis*, and *Cyclocheilichthys repasson* revealed their potential to acclimate and develop the flock in the impoundment condition. Meanwhile, the obligatory riverine species such as *Cirrhinus jullieni*, *Dangila siamensis*, *Notopterus notopterus*, and *Labeo chrysophekadion* showed a continuous decline in fish landing after post-impoundment. On the other hand, species such as *Channa striata*, *Hemibagrus nemurus*, *Mastacembelus armatus*, *Barbonymus* spp., and *Oxyeleotris marmorata* increased in fish landings after a certain period of post-impoundment. Many studies had been done to investigate biological traits and population dynamics of the riverine origin species that are well established and become the dominant in catches in the reservoirs. However, when compared to the same species that inhabit in the river, it was found that these fishes showed the delay in maturity, slower growth, and shorter in asymptotic length (e.g., Jutagate et al. 2003; Suvarnaraksha et al. 2011).

The commonest mitigation of this damming type is fish stocking program which is regularly employed in Thai reservoirs for the main purposes of creation and enhancement the fisheries (Jutagate and Rattanachai 2010). Since 1980, the stocking of indigenous species was inaugurated to be released to the reservoirs countrywide replacing the common practice on stocking of Chinese and Indian carps as well as tilapias (Pawaputanon 1986). The popular species for stocking are *Barbonymus gonionotus*, *Probarbus jullieni*, *Clarias macrocephalus*, *H. siamensis*, *Pangasius hypophthalmus*, *Barbodes schwanenfeldii*, *Barbonymus altus*, *Pangasius larnaudii*, and *Thynnichthys thynnoides* (Virapat et al. 2000). Moreover, stocking to retain the endangered and threatened species is also practiced by the Department of Fisheries such as stocking of *Pangasius sanitwongsei* and *Pangasianodon gigas* (Hogan et al. 2008).

"Anti-salt" Damming to Fish Community

This type of damming is common in the southern part of the country, where it is on the peninsula located between the Gulf of Thailand on the right and Andaman Sea on the left. The main purpose of the dam type is to prevent the intrusion of seawater upstream, especially during the dry season, as in the case in Pak Phanang River (Nakhon Si Thammarat Province) and Bang Nara River (Pattani Province), where the seawater can move as far as more than three-quarters of the total length of the river during the dry season (Prabnarong and Kaewrat 2006; Jutagate et al. 2010). The regulation scheme of this damming type is that the sluice gates are closed during the dry season and occasionally opened during the wet season, depending on the upstream water level, which also allowed fishes to move between up- and downstreams.

According to the regulation, changes in the fish assemblages were found, both in the estuary and in the river. Species richness and the diversity index did not differ in the estuary between the closing and opening phases. However, there were significantly differences in the river in both parameters, i.e., higher during the opening phase. Moreover, salinity in the estuary significantly declined, but significantly increased in the river during the opening phase (Jutagate et al. 2010). From the study in Pak Phanang River, during the opening phase, euryhaline species such as *Osteogeneiosus militaris*, *Leiognathus* spp., and *Mystus gulio* contributed more to the catches in the freshwater area. However, none freshwater fish species was common in the estuarine/marine area (Jutagate et al. 2010, 2011). Meanwhile, the performance of the fish ladder at the dam site is poor due to the inappropriate design, which few species can utilize the ladder to move between the two systems (Rattanavinitkul et al. 2011).

The major fish assemblage patterns, in the estuarine system, were classified as:

(a) The pattern that is dominated by the stenohaline species, which are abundant in the lower saline area such as *Trypauchen vagina*, *Parapocryptes serperaster*, *Platycephalus indicus*, and *Siganus canaliculatus*.
(b) The pattern that is dominated by the true brackish water species, which are the permanent residents in the estuary with euryhaline characteristics, such as *Hilsa kelee*, *Ambassis gymnocephalus*, *Encrasicholina devisi*, *M. gulio*, and *Liza subviridis*, which can enter to the river when the sluice gates are opened.
(c) The pattern that is dominated by the opportunistic marine fish, which sometimes enter the estuary for feeding and breeding purposes, such as *Pampus argenteus*, *Megalops cyprinoides*, *Sardinella gibbosa*, *Rastrelliger brachysoma*, and *Terapon jarbua*.

The assemblage patterns in the river portion can be explained by the riverine fish environmental guilds (see Welcomme et al. 2006 for detail). The fish assemblages were arrayed along a longitudinal gradient, where salinity was also important controlling variable. There were two main assemblage patterns in this portion, i.e., assemblage patterns of (a) upstream and (b) downstream of the dam. The upstream assemblage comprised fishes in lowland river (i.e., potamonic) guilds both lentic species (e.g., *Channa* spp. and *Oxyeleotris marmorata*) and lotic species (e.g., *Notopterus notopterus* and cyprinid fishes). As the assemblage downstream, there are fishes in the guilds of freshwater estuarine (e.g., *Osteogeneiosus militaris* and *Papilogobius reichei*), brackish water estuarine (e.g., *A. gymnocephalus* and *Stolephorus dubiosus*), and amphidromous fishes (e.g., *Megalops cyprinoids* and *Muraenesox cinereus*).

Conclusions

Dams alter river ecosystems and fish as a major component is inevitably affected, in particular blockage of fish migration and losses of the complexity of riverine species richness. Appropriate mitigations are needed to balance the infrastructure development and integrity of the component ecosystem. For example, the common mitigation as a fish passage system has been mostly designed for North America and Europe where the number of fish species is limited, which cannot cope with the rivers with high diversity as in Thailand. Moreover, understanding on biological traits of the riverine species is also crucially needed to determine which species are likely to adapt to the new environment and which species are extirpated.

Acknowledgments We thank the National Institute for Environmental Studies (NIES), Japan, for the financial support and Dr. Michio Fukushima (NIES) for commenting on the manuscript.

References

Baran E, Myschowoda C (2009) Dams and fisheries in the Mekong Basin. Aquat Ecosyst Health Manag 12(3):227–234

De Silva SS, Amarasinghe US (2009) Status of reservoir fisheries in five Asian countries. NACA Monograph No. 2. Network of Aquaculture Centres in Asia-Pacific, Bangkok

Dugan PJ, Barlow C, Agostino AA, Baran E, Cada GF, Chen D, Cowx IG, Ferguson JW, Jutagate T, Mallen-Cooper M, Marmulla G, Nestler J, Petrere M, Welcomme RL, Winemiller KO (2010) Fish migration, dams, and loss of ecosystem services in the Mekong Basin. Ambio 39:344–348

Froese F, Pauly D (2012) FishBase. www.fishbase.org. Accessed 23 Dec

Hogan Z, Na-Nakorn U, Kong H (2008) Threatened fishes of the world: *Pangasius sanitwongsei* Smith 1931 (Siluriformes: Pangasiidae). Environ Biol Fishes 84(3):305–306

Hortle KG, Bamrungrach P (2012) Wetlands, flooding and fisheries in the lower Mekong Basin. Mekong River Commission, Vientiane

Jutagate T, Lamkom T, Satapornwanit K, Naiwinit W, Petchuay C (2001) Fish species diversity and ichthyomass in Pak Mun Reservoir, Thailand, five years after impoundment. Asian Fish Sci 14:417–425

Jutagate T, Krudphan C (2004) Population dynamics of the black shark minnow Morulius chrysophekadion (Cyprinidae) in the run-of-the river in Thailand. Asian Fish Sci 17:1–8

Jutagate T, Krudpan C, Ngamsnae P, Lamkom T, Payooha K (2005) Changes in the fish catches during the trial of opening the sluice gates of a run-of-the river reservoir in Thailand. Fish Manag Ecol 12:57–62

Jutagate T, Thappanand T, Tabthipwan P (2007) Is the sluice gates' management beneficial for spawning migration? The case of shark catfish (Helicophagus waandersii) in the Mun below Pak Mun Dam, Thailand. River Res Appl 23:87–97

Jutagate T, Poomkong P, Rattanajamnong D, Kwangkhang W, Lek S (2008) Mitigation measure helps conserve the fish biodiversity from river damming by applying sluice gate management scheme. Paper presented at The international symposium "Sustaining fish diversity, fisheries and aquacultures in the Mekong Basin", 3–5 Sept 2008. Ubon Ratchathni, Thailand

Jutagate T (2009) Reservoir fisheries of Thailand. In: SS de Silva, Amarasinghe U.S. (eds) Status of reservoir fisheries in five Asian countries. NACA monograph no. 2. Network of Aquaculture Centres in Asia-Pacific, Bangkok

Jutagate T, De Silva SS, Mattson NS (2003) Yield, growth and mortality rate of the Thai river sprat (*Clupeichthys aesarnensis* Wongratana, 1983) in Sirinthorn Reservoir. Thailand Fish Manag Ecol 10:221–231

Jutagate T, Rattanachai A (2010) Inland fisheries resource enhancement and conservation in Thailand. In: De Silva SS, Davy B, Wiemian M (eds) Inland fisheries resource enhancement and conservation in Asia-Pacific. FAO/RAP Publication 2010/22. FAO/RAP, Bangkok

Jutagate T, Swusdee A, Thappanand-Chaidee T, Lek S, Grenouillet G, Thongkhoa S, Chotipuntu P (2010) Variations of environmental variables and fish assemblages due to damming for anti-salt intrusion to the upriver in the tropic. Mar Freshw Res 61(3):288–301

Jutagate T, Lek S, Swusdee A, Sukdiseth U, Thappanand-Chaidee T, Ang-Lek S, Thongkhoa S, Chotipuntu P (2011) Spatio-temporal variations in fish assemblages in a tropical regulated lower river course: an environmental guild approach. River Res Appl 27(1):47–58

Jutagate T, Srichareondham B, Lek S, Amaraasinghe US, De Silva SS (2012) Variations, trends and patterns in fish yields of large reservoirs in Thailand. Lakes Reservoirs: Res Manag 17 (1):35–53

Larinier M (2001) Environmental issues, dams and fish migration. In: Marmulla G (ed) Dams, fish and fisheries: opportunities, challenges and conflict resolution, vol 419, FAO Fish. Tech Pap. FAO, Rome

Marmulla G (2001) Dams, fish and fisheries: opportunities, challenges and conflict resolution, vol 419, FAO Fish. Tech Pap. FAO, Rome

Nguyen TTT, De Silva SS (2006) Freshwater finfish biodiversity and conservation: an Asian perspective. Biodivers Conserv 15:3543–3568

Office of Natural Resources and Environmental Policy and Planning (2013) Red-data of Thailand. http://chm-thai.onep.go.th/RedData/. Access 24 Feb

Pawaputanon O (1986) Fisheries and fishery management of large reservoir in Thailand. In: McClean JL, Dizon DB, Hosillos LV (eds) Proceedings of the first Asian fisheries forum. Asian Fisheries Society, Manila

Prabnarong P, Kaewrat J (2006) The Uthokawiphatprasit watergate: a man made change in Pak Panang River Basin. Walailak J Sci Technol 3:131–143

Raja AK, Srivastava AP, Dwivedi M (2006) Power plant engineering. New Age International, New Delhi

Rattanavinitkul S, Phumchuai C, Nudee V, Kumnerdmanee M, Chomboon C, Kongjan J (2011) Aquatic animals found at the fish ladder of the Uthokawiphatprasit watergate during 2006–2010. Advisory Article No. 1/2011. Coastal Fisheries Research and Development Bureau, Department of Fisheries, Bangkok (in Thai)

Roberts TR (1995) Mekong mainstream hydropower dams: run-of-the-river or ruin-of-the-river? Nat Hist Bull Siam Soc 43(1):9–20

Roberts TR (2001) On the river of no returns: Thailand's Pak Mun Dam and its fish ladder. Nat Hist Bull Siam Soc 49:189–230

Suvarnaraksha A, Lek S, Lek-Ang S, Jutagate T (2011) The life history of the riverine cyprinid *Henicorhynchus siamensis* (Sauvage, 1881) in a small reservoir. J Appl Ichthyol 27 (4):995–1000

Suvarnaraksha A, Lek S, Lek-Ang S, Jutagate T (2012) Fish diversity and assemblage patterns along the longitudinal gradient of tropical river in the Indo-Burma hotspot region (the Ping-Wang River Basin, Thailand). Hydrobiol 694:153–169

Trussart S, Messier D, Roquet V, Aki S (2002) Hydropower projects: a review of most effective mitigation measures. Energ Policy 30:1251–1259

Virapat C, Phimonbutra U, Chantarawaratid C (2000) Fishery and fisheries management in Thai reservoirs: review and assessment. Mekong River Commission, Vientiane

Welcomme RL, Winemiller KO, Cowx IG (2006) Fish environmental guilds as a tool for assessment of ecological condition of rivers. River Res Appl 22:377–396

Chapter 4
Potential Future Coral Habitats Around Japan Depend Strongly on Anthropogenic CO_2 Emissions

Yumiko Yara, Hiroya Yamano, Marco Steinacher, Masahiko Fujii, Meike Vogt, Nicolas Gruber, and Yasuhiro Yamanaka

Abstract Using the results from the NCAR CSM1.4-coupled global carbon cycle–climate model under the Intergovernmental Panel on Climate Change (IPCC) emission scenarios SRES A2 and B1, we estimated the effects of both global warming and ocean acidification on the future habitats of corals in the seas around Japan during this century. As shown by Yara et al. (Biogeosciences 9:4955–4968, 2012), under the high-CO_2-emission scenario (SRES A2), coral habitats will be sandwiched and narrowed between the northern region, where the saturation state of the carbonate mineral aragonite (Ω_{arag}) decreases, and the southern region, where coral bleaching occurs. We found that under the low-emission scenario SRES B1, the coral habitats will also shrink in the northern region by the reduced Ω_{arag} but to a lesser extent than under SRES A2, and in contrast to SRES A2, no bleaching will occur in the southern region. Therefore, coral habitats in the southern region are

Y. Yara
Center for Environmental Biology and Ecosystem Studies, National Institute for Environmental Studies, 16-2 Onogawa, Tsukuba, Ibaraki 305-8506, Japan

Project Team for Analyses of Changes in East Japan Marine Ecosystems, JAMSTEC, 2-15 Natsushima, Yokosuka, Kanagawa 237-0061, Japan

H. Yamano (✉)
Center for Environmental Biology and Ecosystem Studies, National Institute for Environmental Studies, 16-2 Onogawa, Tsukuba, Ibaraki 305-8506, Japan
e-mail: hyamano@nies.go.jp

M. Steinacher
Climate and Environmental Physics, Physics Institute, University of Bern, Bern, Switzerland

M. Fujii • Y. Yamanaka
Faculty of Environmental Earth Science, Hokkaido University, N10W5, Kita-ku, Sapporo, Hokkaido 060-0810, Japan

M. Vogt • N. Gruber
Environmental Physics Group, Institute of Biogeochemistry and Pollutant Dynamics, ETH Zurich, Zurich, Switzerland

S.-i. Nakano et al. (eds.), *Aquatic Biodiversity Conservation and Ecosystem Services*, Ecological Research Monographs, DOI 10.1007/978-981-10-0780-4_4

expected to be largely unaffected by ocean acidification or surface warming under the low-emission scenario. Our results show that potential future coral habitats depend strongly on CO_2 emissions and emphasize the importance of reducing CO_2 emissions to prevent negative impacts on coral habitats.

Keywords CO_2 emission scenarios • Climate change • Global warming • Ocean acidification • Coral • Japan

Introduction

Anthropogenic CO_2 emissions have caused atmospheric CO_2 to rise by more than 30 % since the preindustrial era. This has changed the oceanic environment in at least two ways. First, it caused an increase in the global average sea surface temperature (SST) of about 0.5 °C in response to the global warming induced by the increase in atmospheric CO_2 and in the concentration of other greenhouse gases (Trenberth et al. 2007). Second, it caused a reduction in the surface ocean pH of about 0.1 units and a reduction in the saturation state of the carbonate mineral aragonite (Ω_{arag}) of about 0.3 units (Feely et al. 2004). If CO_2 emissions continue to increase, much larger changes, including increases in SST of several degrees Celsius, reductions in pH of up to nearly one unit, and even more severe reductions in Ω_{arag}, will occur in the future (e.g., Gruber 2011; Feely et al. 2009; Bopp et al. 2013).

These changes may affect the distribution of coral habitats. Increases in SST in response to global warming will allow coral habitats to expand poleward in temperate areas (Precht and Aronson 2004; Yamano et al. 2011), while at the same time, possibly reducing the present coral habitats by excessive coral bleaching in subtropical/tropical areas (Guinotte et al. 2003; Meissner et al. 2012; Hoegh-Guldberg et al. 2007; Frieler et al. 2013). Lowered Ω_{arag} caused by ocean acidification will affect the life histories of corals (e.g., Gattuso et al. 1998; Kleypas et al. 1999a; Langdon et al. 2000, 2003; Anthony et al. 2008, 2011; Morita et al. 2009; Albright et al. 2010), particularly at higher latitudes where Ω_{arag} will reach critically low levels earlier than at low latitudes because the SSTs are lower (e.g., Kleypas et al. 1999b; Orr et al. 2005; Yamamoto-Kawai et al. 2009; Steinacher et al. 2009). Emerging field evidence at CO_2 seepage sites suggests that acidified waters do not allow coral survival (Fabricius et al. 2011; Inoue et al. 2013).

Until recently, the impacts of ocean acidification and global warming on future coral habitats have been examined on a global scale, without considering the shifts in their distributional ranges (Kleypas et al. 1999b; Guinotte et al. 2003; Meissner et al. 2012; Frieler et al. 2013; Hooidonk et al. 2013). Under the high-CO_2-emission SRES A2 scenario (Fig. 4.1), Yara et al. (2012) projected that in Japan, which is located at the northern limit of coral and coral reef distributions, coral habitats will decrease from the north because of ocean acidification, although corals are

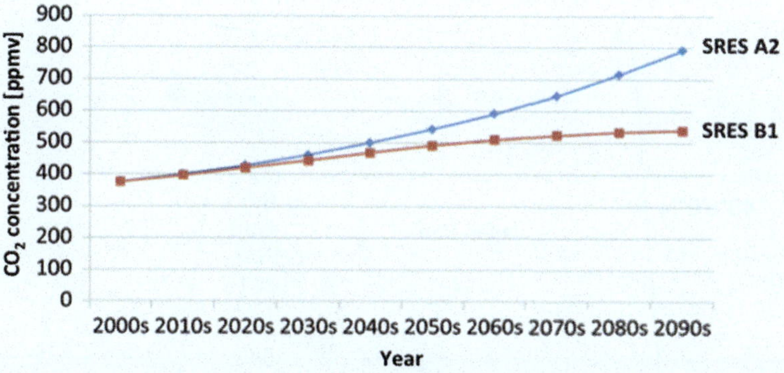

Fig. 4.1 Anthropogenic CO_2 emission scenarios prepared for the Fourth Assessment Report of the Intergovernmental Panel on Climate Change (Trenberth et al. 2007)

projected to expand northward in response to SST warming. Furthermore, SST warming in the southern region will cause coral bleaching. Therefore, under the SRES A2 scenario, the coral habitats around Japan will be lost by the end of the twenty-first century. But while it is clear that a lower-emission scenario will cause a much smaller contraction of the habitats (Gattuso et al. 2015), the response is not linear. Thus it is critical that these phenomena must also be assessed under scenarios of lower CO_2 emissions, to examine the dependence of future coral habitats on CO_2 emissions.

In this study, using both coarse-resolution temporal–spatial data obtained from a climate system model and the simplified indices of SST and Ω_{arag} developed by Yara et al. (2012), we estimated the effects of both global warming and ocean acidification on the potential future habitats of corals under the high-CO_2-emission scenario (SRES A2) and the low-CO_2-emission scenario (SRES B1, Fig. 4.1) in the seas around Japan during the twenty-first century.

Methods

Study Area

Japan offers a unique opportunity to study the effects of global warming and ocean acidification on corals because the archipelago covers a wide latitudinal range, stretching from subtropical to temperate areas in the East China Sea, the Japan Sea, and the Pacific Ocean (Fig. 4.2). Furthermore, the distribution of corals around the Japanese islands has been described in detail (e.g., Veron and Minchin 1992; Yamano et al. 2011), allowing researchers to predict future changes against the well-established present state.

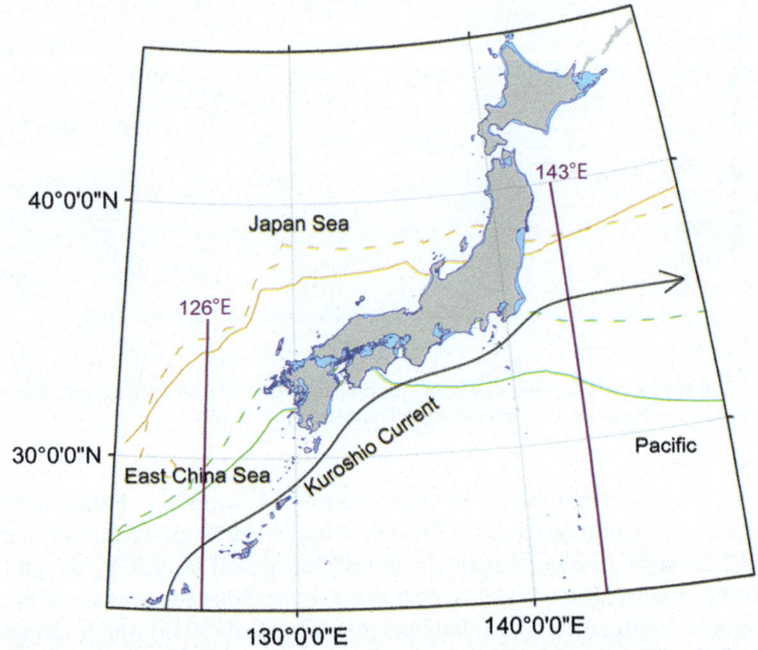

Fig. 4.2 Study site locations. The *violet lines* denote the north–south (*latitudinal*) axis of the East China Sea/Japan Sea (126° E) and Pacific Ocean (143° E) from which we estimated the rates of change in SST and Ω_{arag}. Isolines SST = 18 °C (*solid green line*) and SST = 10 °C (*solid orange line*) of the 20-year mean SST during the coldest month and isolines $\Omega_{arag} = 3$ (*dotted green line*) and $\Omega_{arag} = 2.3$ (*dotted orange line*) of the 20-year mean Ω_{arag} during the annual lowest Ω_{arag}, for each year in 1980–1999 (present), are also presented. Blue regions show areas shallower than 50 m

Future Projection Data Under Two Different Future CO_2 Emission Scenarios

We used future projection data from 1980 to 2099 under SRES A2 and B1 from the Climate System Model CSM1.4-carbon of the US National Center for Atmospheric Research (NCAR; Fung et al. 2005; Doney et al. 2006; Steinacher et al. 2009). Under SRES A2 and B1, atmospheric CO_2 will reach about 828 and 540 ppm, respectively, by the end of the century. The horizontal resolution of the CSM1.4 model is 3.6° × 0.8°–1.8°, and the model data were interpolated onto a regular 1° × 1° grid to match the spatial resolution of the observational data for the western North Pacific (the region bounded by 24°–48° N, 118°–157° E).

To produce the future monthly projection data for SST and Ω_{arag} during this century under SRES A2 and B1, we added the modeled annual mean SST and Ω_{arag} anomalies for 2000–2099 to the monthly observation-based climatologies, following Yara et al. (2012), as described below. We assumed cyclostationary seasonal cycles for SST and Ω_{arag} during this century, because the modeled SST and Ω_{arag} are annual mean data. The modeled anomalies were computed by subtracting the modeled climatological 20-year means for SST and Ω_{arag} over 1980–1999 from the modeled time series for SST and Ω_{arag}. From the seasonal envelope, we derived SST during the hottest/coldest months and the annual lowest Ω_{arag} for each year. Finally, we calculated the decadal means of the bias-corrected SST for the hottest/coldest months and the annual lowest Ω_{arag} for each decade between 2000 and 2099. The monthly observation-based climatologies and the SST and Ω_{arag} data predicted with the NCAR CSM1.4 model under SRES A2 were obtained from Yara et al. (2012). The SST and Ω_{arag} data predicted with the NCAR CSM1.4 model under SRES B1 were obtained from Steinacher et al. (2009). Whereas Yara et al. (2012) analyzed these data with the Grid Analysis and Display System (GrADS) (http://iges.org/grads/), in this study, we analyzed them with ArcGIS (http://www.esri.com/software/arcgis). Although the analytical results under SRES A2 in this study are qualitatively the same as those of Yara et al. (2012), the quantitative differences are based on the different software used.

Simplified Indices for the Northern and Southern Limits of Coral Habitats

These thresholds for SST and Ω_{arag}, used to describe the potential coral habitats, are the same as those used by Yara et al. (2012) and reflect the current distributional limits of tropical/subtropical coral reefs and temperate corals in the seas around Japan. For the tropical/subtropical (reef-building) coral community, we used as the northern limit the isolines SST = 18 °C (Hori 1980; Yara et al. 2012) and $\Omega_{arag} = 3$ (Kleypas et al. 1999a, b, 2006; Guinotte et al. 2003; Yara et al. 2012) during the annual lowest Ω_{arag} for each year at the sea surface. For the temperate coral communities without major reef building, we used as the northern limit the isolines SST = 10 °C (Honma and Kitami 1978; Yara et al. 2012) and $\Omega_{arag} = 2.3$ (Yara et al. 2012) during the annual lowest Ω_{arag} of each year at the sea surface. For the entire coral community, we used the 30 °C isotherm during the hottest months of each year as a southern limit, i.e., the northern limit of the regions in which coral bleaching occurs (Kayanne et al. 1999; Yara et al. 2012).

These thresholds, which define the suitability for a coral habitat, are approximations because experimental data are sparse, and they should be used as baselines to document changes in SST and Ω_{arag} rather than as absolute thresholds. Therefore, we did not consider the potential for acclimation or adaptation to global warming and ocean acidification, as was considered in Yara et al. (2012).

Quantifying the Changes in Coral Habitats

To present the differences between the two emission scenarios quantitatively, we estimated the rates of the latitudinal shifts in the potential coral habitats under each scenario using the simplified habitat suitability indices of isolines that represent the northern and southern boundaries of coral habitats.

We averaged SST and Ω_{arag} longitudinally and projected the progression of the critical Ω_{arag} and SST isolines onto the north–south (latitudinal) axis of the East China Sea/Japan Sea (126° E, hereafter referred to as "the East China Sea/Japan Sea line") and the Pacific Ocean (143° E, hereafter referred to as "the Pacific line") to estimate the rate of change in these variables. The rates differ greatly north and south of the Kuroshio Current, tending to be faster in the south than in the north (Yara et al. 2012). Therefore, the East China Sea/Japan Sea line was established for the seas around the Ryukyu Islands located south of the Kuroshio Current, and the Pacific line was established for the sea around Honshu, located north of the Kuroshio Current. Both the East China Sea/Japan Sea and Pacific lines are shown in Fig. 4.2.

Together with the rates of these shifts, we estimated the changes in the potential coral habitats in response to SST warming and ocean acidification. We regarded the coral habitats as areas shallower than 50 m. Isobaths of 50 m were extracted from the bathymetry data (http://www.jha.or.jp/jp/shop/products/btdd/) prepared by the Japan Coast Guard. We used the threshold values as described above (Yara et al. 2012) for areas shallower than 50 m, to extract the potential coral habitats (Fig. 4.2).

Results and Discussion

The intensities of global warming and ocean acidification, and therefore the future of coral habitats, depend strongly on the CO_2 emission scenario. The differences in the predicted effects of these scenarios on coral habitats were evaluated based on differences between SRES A2 and B1 (Fig. 4.1).

Differences in the Projected Effects of Global Warming on Coral Habitats Under Different Future CO_2 Emission Scenarios

In response to the rising sea temperatures caused by global warming, coral habitats and the regions in which coral bleaching occurs are both projected to expand poleward from the equator. Therefore, the isolines SST = 18 °C (defined as the northern limit of tropical/subtropical coral communities) and SST = 10 °C (defined

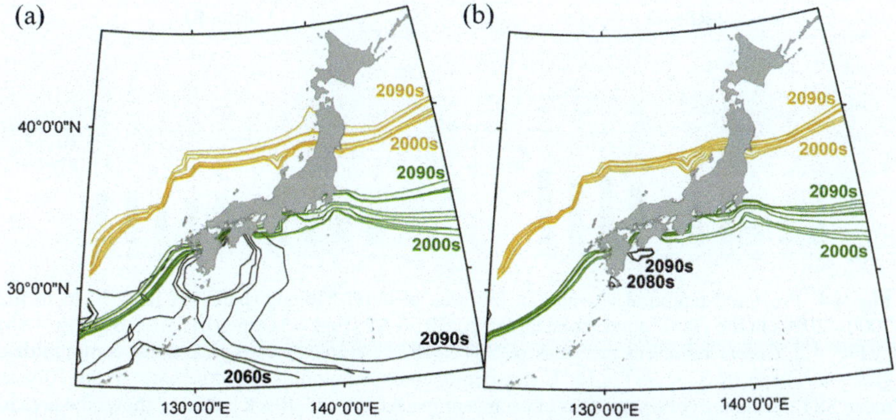

Fig. 4.3 Projected isolines for SST = 18 °C (defined as the northern limit of tropical/subtropical coral communities, *green lines*) and SST = 10 °C (defined as the northern limit of temperate coral communities, *orange lines*) of the 10-year mean SST during the coldest months of each year and SST = 30 °C (defined as the northern limit of regions in which coral bleaching occurs, *black lines*) of the 10-year mean SST during the hottest months of each year in each decade in the 2000s–2090s under (**a**) SRES A2 and (**b**) SRES B1

as the northern limit of temperate coral communities) of the 10-year mean SST during the coldest months of each year, and SST = 30 °C (defined as the northern limit of the regions in which coral bleaching occurs; the southern limits of corals) of the 10-year mean SST during the hottest months of each year in each decade during 2000s–2090s, will move progressively northward in the seas around Japan in response to global warming (Fig. 4.3). The isothermal lines SST = 18 °C and SST = 10 °C differ only slightly under SRES A2 and B1, whereas the SST isothermal line SST = 30 °C differs greatly under SRES A2 and B1. The isothermal line SST = 30 °C enters the area from the south after the 2060s under SRES A2, but only approaches the center of the area after the 2080s under SRES B1. Therefore, unlike SRES A2, coral habitats are not projected to narrow in response to coral bleaching during global warming under SRES B1 in this century in the seas around Japan.

Under SRES A2, the isolines SST = 18 °C and SST = 10 °C will move northward at latitudinal velocities of 1.22 and 0.76 km $year^{-1}$, respectively, onto the East China Sea/Japan Sea line, and at velocities of 2.16 and 1.56 km $year^{-1}$, respectively, onto the Pacific line, and the isoline SST = 30 °C will move northward at a latitudinal velocity of 12.5 km $year^{-1}$ onto the East China Sea/Japan Sea line (Fig. 4.3a). Under SRES B1, the isolines SST = 18 °C and SST = 10 °C will move northward at latitudinal velocities of 0.58 and 0.34 km $year^{-1}$, respectively, onto the East China Sea/Japan Sea line, and at velocities of 1.48 and 0.79 km $year^{-1}$, respectively, onto the Pacific line (Fig. 4.3b). These rate calculations show quantitatively that the rate of expansion is slower under SRES B1 than that under SRES A2.

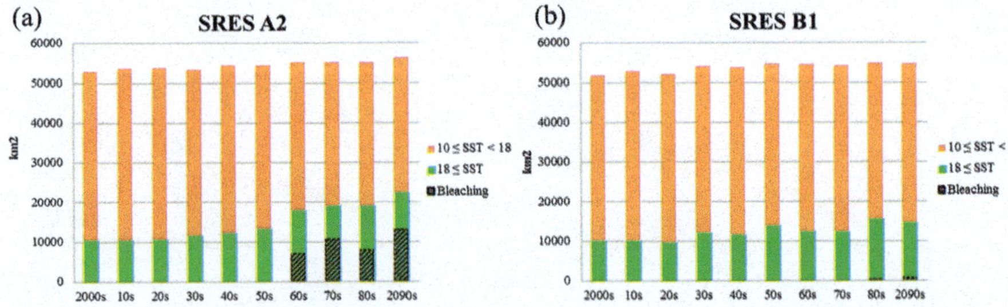

Fig. 4.4 Projected temporal changes in the area of coral habitats shallower than 50 m in the 2000s–2090s in the seas around Japan under (**a**) SRES A2 (Fig. 4.3a) and (**b**) SRES B1 (Fig. 4.3b) in Fig. 4.2. *Colors* represent the areas of SST ≥18 °C (tropical/subtropical coral communities, *green bars*) and 10 °C ≤ SST <18 °C (temperate coral communities, *orange bars*) of the 10-year mean SST during the coldest months of each year in each decade. Black hatching shows the area in which SST ≥30 °C (where coral bleaching occurs) of the 10-year mean SST during the hottest months of each year in each decade

Figure 4.4 presents the changes in the area of coral habitats shallower than 50 m in the 2000s–2090s under (a) SRES A2 (Fig. 4.4a) and (b) SRES B1 (Fig. 4.4b) in Fig. 4.2. The total areas of the coral habitats under the two scenarios differ only slightly during this century in the seas around Japan, and the total area will be 1.07 times larger in the 2090s than in the 2000s under SRES A2 and 1.06 times larger under SRES B1. As is shown by the breakdown of the total area into SST ≥ 18 °C and 10 °C ≤ SST < 18 °C, these areas tend to increase and decrease, respectively, with time under the two scenarios. In contrast, the regions in which coral bleaching occurs will account for 22.8 % of the total area in the 2090s under SRES A2 and for 0.736 % of the total area in the 2090s under SRES B1.

Differences in the Projected Effects of Ocean Acidification on Coral Habitats Under Different Future CO_2 Emission Scenarios

In response to the decreasing Ω_{arag} caused by ocean acidification, the area of lower Ω_{arag}, unsuitable for coral habitat, which will reduce the calcification rates and overall reproductive success of corals, is projected to expand toward the equator. In other words, current coral habitats are projected to shift and narrow toward the equator. Therefore, the isolines $\Omega_{arag} = 3$ (defined as the northern limit of tropical/subtropical coral communities) and $\Omega_{arag} = 2.3$ (defined as the northern limit of temperate coral communities) from the 10-year mean Ω_{arag} during the annual lowest Ω_{arag} of each year in each decade in the 2000s–2090s are moving progressively southward in the seas around Japan with increasing ocean acidification (Fig. 4.5). The isoline $\Omega_{arag} = 3$ disappears after the 2030s under SRES A2 and after the 2040s under B1. In contrast, isoline $\Omega_{arag} = 2.3$ disappears after the 2080s

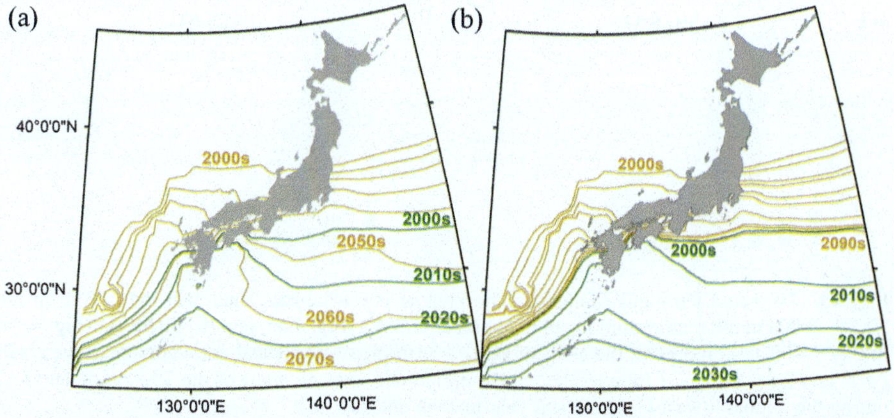

Fig. 4.5 Projected isolines $\Omega_{arag} = 3$ (defined as the northern limit of tropical/subtropical coral communities, *green lines*) and $\Omega_{arag} = 2.3$ (defined as the northern limit of temperate coral communities, *orange lines*) of the 10-year mean Ω_{arag} during the annual lowest Ω_{arag} of each year in each decade of the 2000s–2090s under (**a**) SRES A2 and (**b**) SRES B1

under SRES A2 but does not disappear because it barely shifts after the 2070s under SRES B1. Therefore, based on ocean acidification, coral habitats are projected to remain in the seas around Japan throughout this century under SRES B1, unlike under SRES A2.

Under SRES A2, the $\Omega_{arag} = 3$ isoline located south of the Kuroshio Current moves southward at a latitudinal velocity of 11.5 km year^{-1}, onto the East China Sea/Japan Sea line, and moves southward at a latitudinal velocity of 34.9 km year^{-1}, onto the Pacific line (Fig. 4.4a). The speed of the $\Omega_{arag} = 2.3$ isoline was evaluated in the regions north and south of the Kuroshio Current, because the $\Omega_{arag} = 2.3$ isoline crosses the Kuroshio Current. Across the regions north and south of the Kuroshio Current, the latitudinal Ω_{arag} gradient is small (as is the SST gradient), so slight differences in the projected Ω_{arag} (and SST) lead to large differences in the locations of the isolines with time (Yara et al. 2012). The $\Omega_{arag} = 2.3$ isoline located north and south of the Kuroshio Current will move southward at a latitudinal velocity of 14.0 km year^{-1} in the 2000s–2040s and 14.2 km year^{-1} in the 2040s–2070s, onto the East China Sea/Japan Sea line, and will move southward at a latitudinal velocity of 9.03 km year^{-1} in the 2000s–2040s and 34.3 km year^{-1} in the 2040s–2070s, onto the Pacific line (Fig. 4.5a). Under SRES B1, the isolines $\Omega_{arag} = 3$ and $\Omega_{arag} = 2.3$ will move southward at latitudinal velocities of 11.0 and 6.98 km year^{-1}, respectively, onto the East China Sea/Japan Sea line, and southward at latitudinal velocities of 27.7 and 5.43 km year^{-1}, respectively, onto the Pacific line (Fig. 4.5b). These speed calculations show quantitatively that the rate of expansion under SRES B1 is slower than that under SRES A2.

Figure 4.6 presents the changes in the areas of coral habitats shallower than 50 m in the 2000s–2090s under (a) SRES A2 (Fig. 4.6a) and (b) SRES B1 (Fig. 4.6b) in Fig. 4.4. The total areas of coral habitat under the two scenarios differ greatly after

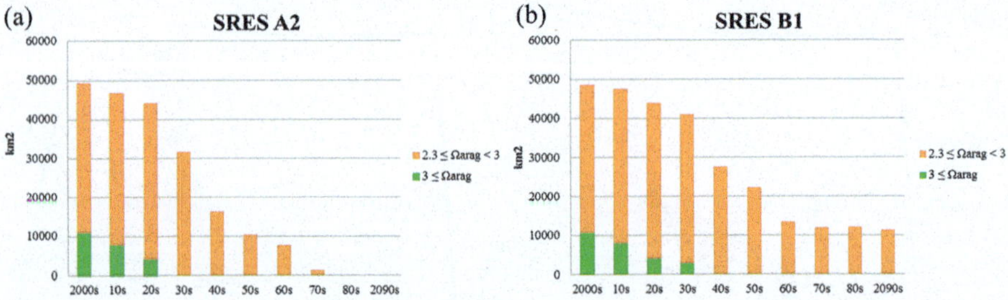

Fig. 4.6 Projected temporal changes in the area of coral habitats shallower than 50 m in the 2000s–2090s in the seas around Japan under (**a**) SRES A2 (Fig. 4.5a) and (**b**) SRES B1 (Fig. 4.5b) in Fig. 4.4. *Colors* represent the areas of Ω_{arag} ≥3 (tropical/subtropical coral communities, *green bars*) and 2.3≤ Ω_{arag} <3 (temperate coral communities, *orange bars*) of the 10-year mean Ω_{arag} during the annual lowest Ω_{arag} of each year in each decade

the 2030s in the seas around Japan. Under SRES A2, the total area decreases and disappears after the 2080s. In contrast, under SRES B1, the total area decreases more slowly after the 2030s than under SRES A2 and hardly decreases at all after the 2070s, and in the 2090s, the total area is 0.23 times larger than the total area in the 2000s. As shown by the total area breakdown, the area of $\Omega_{arag} \geq 3$ disappears during the first half of this century under both scenarios. However, the area of $2.3 \leq \Omega_{arag} < 3$ disappears during the second half of this century under SRES A2, but does not disappear during this century under SRES B1.

Dependence of Potential Future Coral Habitats Around Japan on CO_2 Emission Scenarios

We conclude that if SST was the only factor regulating coral habitats, species could potentially keep up with the rate of change in SST and the entire coral ecosystem would migrate poleward without major impediments. However, the southward progression of reduced Ω_{arag} will probably limit this expansion for those species sensitive to ocean acidification, such as corals and mollusks (e.g., Albright et al. 2010; Morita et al. 2009).

The results for global warming reported in section “Differences in the projected effects of global warming on coral habitats under different future CO_2 emission scenarios” and the results for ocean acidification reported in section “Differences in the projected effects of ocean acidification on coral habitats under different future CO_2 emission scenarios” overlap in the area of SST ≥10 °C where the ocean depth is <50 m, which corals can inhabit (Fig. 4.7). As coral habitats shift northward, the corals will become increasingly immersed in waters with lower Ω_{arag} because of ocean acidification. The region in which coral bleaching occurs in response to global warming will also increase and is expected to expand from the south. As a result, under SRES A2, the coral habitats will be sandwiched and will narrow

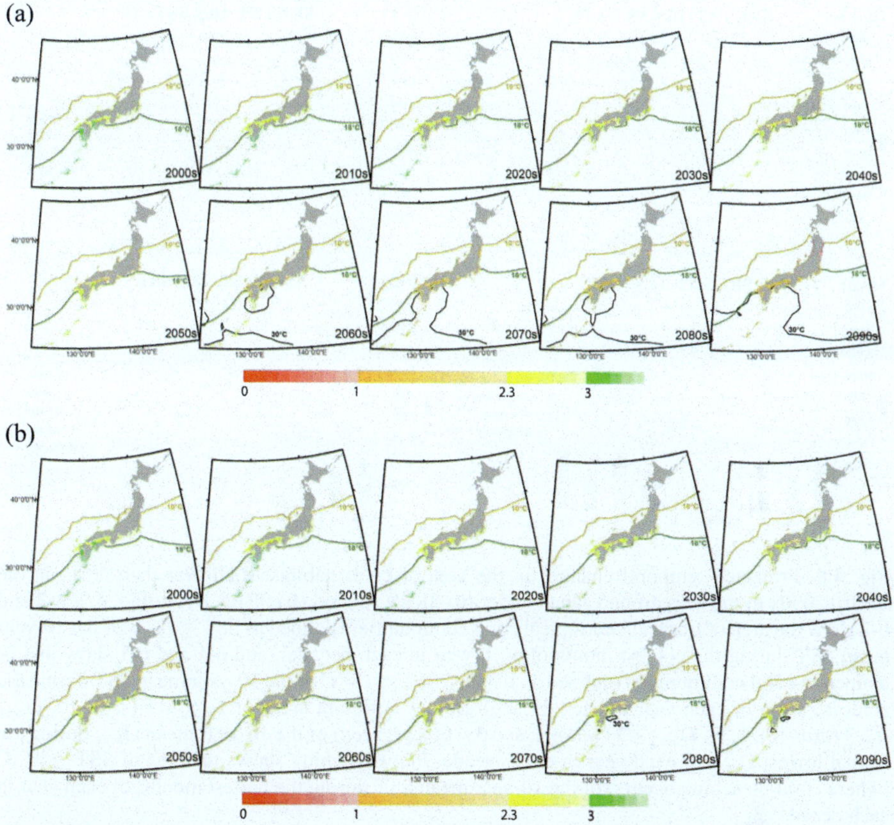

Fig. 4.7 Figures 4.3 and 4.6 were superimposed at SST ≥10 °C below a depth of 50 m in the seas around Japan under (**a**) SRES A2 and (**b**) SRES B1. The *lines* depict global warming, and the *shading* depicts ocean acidification. The *green lines* denote the 10-year mean northern limit of tropical/subtropical coral communities (defined by the isoline SST = 18 °C of the 10-year mean SST during the coldest month of each year in each decade), and the *orange lines* denote that for the temperate coral communities (defined by the isoline SST = 10 °C of the 10-year mean SST during the coldest month of each year in each decade). The *black lines* denote the 10-year mean northern limit of regions in which coral bleaching occurs (defined by the isoline SST = 30 °C of the 10-year mean SST during the hottest month of each year in each decade). *Shading* denotes the 10-year mean Ω_{arag} during the annual lowest Ω_{arag} of each year in each decade

between the northern regions where Ω_{arag} decreases and the southern regions where coral bleaching occurs. However, under SRES B1, coral habitats will only be narrowed by the northern regions in which Ω_{arag} decreases. Therefore, under the two scenarios, the marginal areas occupied by temperate coral communities will remain, whereas suitable/marginal areas for tropical/subtropical coral communities will not remain. However, the areas of tropical/subtropical coral communities could remain under SRES B1 if reefs develop where Ω_{arag} ≥2.3 or can tolerate low Ω_{arag}.

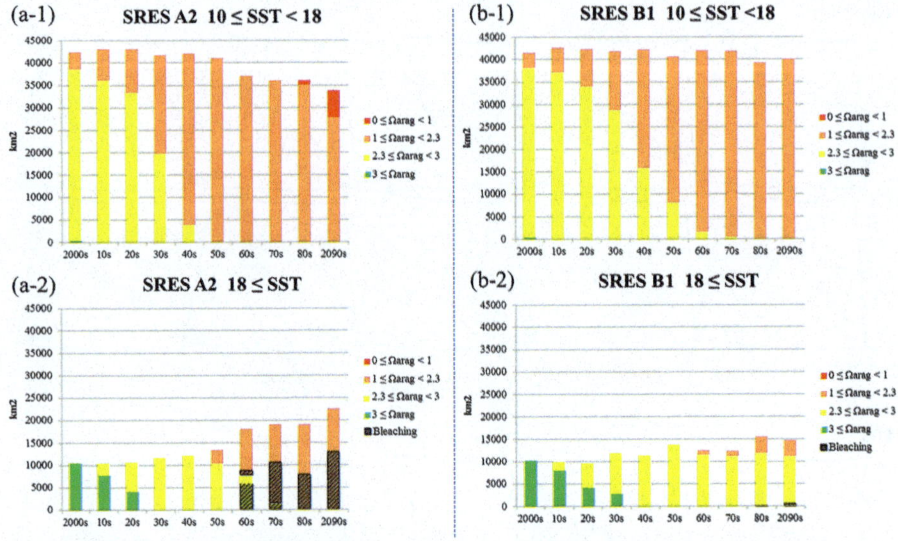

Fig. 4.8 Projected temporal changes in the area of coral habitats shallower than 50 m in the 2000s–2090s in the seas around Japan under (**a**) SRES A2 and (**b**) SRES B1, in Fig. 4.7. **a-2** and **b-2** show the area of tropical/subtropical coral communities (defined as SST ≥18 °C of the 10-year mean SST during the coldest month of each year in each decade), and **a-1** and **b-1** show that of temperate coral communities (defined as 10 °C ≤ SST < 18 °C of the 10-year mean SST during the coldest month of each year in each decade). *Colors* represent areas of Ω_{arag} <1 (*red*), 1≤ Ω_{arag} <2.3 (*orange*), 2.3≤ Ω_{arag} <3 (*yellow*), and 3≤ Ω_{arag} (*green*) of the 10-year mean Ω_{arag} during the annual lowest Ω_{arag} of each year in each decade. *Black* hatching shows the area of SST ≥30 °C (where coral bleaching occurs) of the 10-year mean SST during the hottest months of each year in each decade

Therefore, the extent of future habitats suitable for corals will depend on the CO_2 emission scenario in place.

Figure 4.8 shows the changes in the areas of regions inhabited by tropical/subtropical coral communities (defined as SST ≥18 °C of the 10-year mean SST during the coldest month of each year in each decade) and by temperate coral communities (defined as 10 °C ≤ SST < 18 °C of the 10-year mean SST during the coldest month of each year in each decade) at depths shallower than 50 m, during the 2000s 2090s in the seas around Japan under (a) SRES A2 and (b) SRES B1 (Fig. 4.7). In the regions inhabited by tropical/subtropical coral communities (Fig. 4.8a-2, b-2), the regions of Ω_{arag} ≥3, defined as the marginal area for tropical/subtropical corals based on Yara et al. (2012), constitute almost the total region in the 2000s and are projected to disappear after the 2030s under SRES A2 and after the 2040s under SRES B1. The region of Ω_{arag} <3, defined as unsuitable for tropical/subtropical corals based on Yara et al. (2012), will increase with time. The regions in which coral bleaching occurs (defined as SST ≥30 °C of the 10-year mean SST during the hottest months of each year in each decade) will then increase after the 2060s and will account for 56.8 % of the total area in the 2090s under

SRES A2 but will account for 2.7 % of the total area in the 2090s under SRES B1. In regions inhabited by temperate coral communities (Fig. 4.8a-1, b-1), the regions in which $\Omega_{arag} \geq 2.3$, defined as suitable areas for temperate corals based on Yara et al. (2012), will account for 91.2 % and 92.1 % of the total area in the 2000s under SRES A2 and B1, respectively, but are projected to disappear after the 2050s under SRES A2 and after the 2080s under SRES B1. The regions of $1 \leq \Omega_{arag} < 2.3$, defined as the marginal areas for temperate corals based on Yara et al. (2012), will increase with time. The region of $0 \leq \Omega_{arag} < 1$ (where seawater is undersaturated), defined as unsuitable areas for temperate corals based on Yara et al. (2012), will increase after the 2080s and will account for 17.3 % of the total area in the 2090s under SRES A2.

Conclusion

Based on a coupled global carbon cycle–climate model and simple indices (SST and Ω_{arag}) of coral distributions, we projected the potential future habitats of corals in the seas around Japan during this century under future high- and low-CO_2-emission scenarios (SRES A2 and B1, respectively). We note that our results are based on a single, relatively simple coarse-resolution general circulation model and that other models and approaches have generated somewhat different results when attempting to isolate the impacts of warming and acidification (Couce et al. 2013).

In the seas around Japan, under the high-CO_2-emission scenario (SRES A2), the coral habitats will be sandwiched and narrowed between northern regions in which Ω_{arag} decreases and southern regions in which coral bleaching occurs, whereas under the low-CO_2-emission scenario (SRES B1), the coral habitats will only be narrowed by the northern regions in which Ω_{arag} decreases. Under the low-CO_2-emission scenario, the area of coral will remain if reefs can tolerate low Ω_{arag}. Therefore, the available future coral habitats depend strongly on the levels of CO_2 emissions. In other words, to avoid any potential negative effects on coral reef habitats, stringent reductions in future CO_2 emissions from human activities will be required, as is the case for a range of different climatically affected targets (Steinacher et al. 2013). Our results can be used as a baseline from which to predict and understand the status of corals under different future CO_2 emission scenarios.

Acknowledgments We thank D. Loher for providing the interpolated GLODAP/observational fields. Y. Yara and H. Y. acknowledge funding by the Strategic Research and Development Project (S-9) of the Ministry of the Environment, Japan. H. Y., M. F., and Y. Yamanaka acknowledge support under the auspices of the "Precise Impact Assessments on Climate Change" of the Program for Risk Information on Climate Change (SOUSEI Program) supported by the Ministry of Education, Culture, Sports, Science, and Technology of Japan (MEXT). M. S. acknowledges the support of the Swiss National Science Foundation and the "European Project on Ocean Acidification," EPOCA (211384), and the European Project CARBOCHANGE (264879), which both receive funding from the European Commission's Seventh Framework Programme (FP7/20072013). M. V. and N. G. acknowledge funding from ETH Zurich.

References

Albright R, Mason B, Miller M, Langdon C (2010) Ocean acidification compromises recruitment success of the threatened Caribbean coral *Acropora palmata*. Proc Natl Acad Sci U S A 107: 20400–20404

Anthony KRN, Kline DI, Diaz-Pulido G, Dove S, Hoegh-Guldberg O (2008) Ocean acidification causes bleaching and productivity loss in coral reef builders. Proc Natl Acad Sci U S A 105: 17442–17446

Anthony KRN, Maynard JA, Diaz-Pulido G, Mumby PJ, Marshall PA, Cao L, Hoegh-Guldberg O (2011) Ocean acidification and warming will lower coral reef resilience. Glob Chang Biol 17: 1798–1808

Bopp L, Resplandy L, Orr JC, Doney SC, Dunne JP, Gehlen M, Halloran P, Heinze C, Ilyina T, Séférian R, Tjiputra J, Vichi M (2013) Multiple stressors of ocean ecosystems in the 21st century: projections with CMIP5 models. Biogeosciences 10:6225–6245

Cohen AL, Holcomb M (2009) Why corals care about ocean acidification: uncovering the mechanism. Oceanography 22:118–127

Couce EA, Ridgwell A, Henry EJ (2013) Future habitat suitability for coral reef ecosystems under global warming and ocean acidification. Glob Chang Biol 19:3592–3606

Doney S, Lindsay K, Fung I, John J (2006) Natural variability in a stable, 1000-year global coupled climate-carbon cycle simulation. J Climate 19:3033–3054

Fabricius KE, Langdon C, Uthicke S, Humphrey C, Noonan S, De'ath G, Okazaki R, Muehllehner N, Glas MS, Lough JM (2011) Losers and winners in coral reefs acclimatized to elevated carbon dioxide concentrations. Nat Clim Chang 1:165–169

Feely RA, Sabine CL, Lee K, Berelson W, Kleypas J, Fabry VJ, Millero FJ (2004) Impact of anthropogenic CO_2 on the $CaCO_3$ system in the oceans. Science 305:362–366

Feely RA, Doney SC, Cooley SR (2009) Ocean acidification: present conditions and future changes in a high-CO_2 world. Oceanography 22:36–47

Frieler K, Meinshausen M, Golly A, Mengel M, Lebek K, Donner SD, Hoegh-Guldberg O (2013) Limiting global warming to 2°C is unlikely to save most coral reefs. Nat Clim Chang 3: 165–170

Fung I, Doney S, Lindsay K, John J (2005) Evolution of carbon sinks in a changing climate. Proc Natl Acad Sci U S A 102:11201–11206

Gattuso J-P, Frankignoulle M, Bourge I, Romaine S, Buddemeier RW (1998) Effect of calcium carbonate saturation of seawater on coral calcification. Glob Planet Chang 18:37–46

Gattuso J-P, Magnan A, Billé R, Cheung WWL, Howes EL, Joos F, Allemand D, Bopp L, Cooley SR, Eakin CM, Hoegh-Guldberg O, Kelly RP, Pörtner H-O, Rogers AD, Baxter JM, Laffoley D, Osborn D, Rankovic A, Rochette J, Sumaila UR, Treyer S, Turley C (2015) Contrasting futures for ocean and society from different anthropogenic CO_2 emissions scenarios. Science 349: aac4722

Gruber N (2011) Warming up, turning sour, losing breath: ocean biogeochemistry under global change. Phil Trans R Soc Lond A 369:1980–1996

Gruber N, Hauri C, Lachkar Z, Loher D, Frölicher TL, Plattner G-K (2012) Rapid progression of ocean acidification in the California current system. Science 337:220–223

Guinotte JM, Buddemeier RW, Kleypas JA (2003) Future coral reef habitat marginality: temporal and spatial effects of climate change in the Pacific basin. Coral Reefs 22:551–558

Hoegh-Guldberg O (2005) Low coral cover in a high-CO_2 world. J Geophys Res 110:C09S06

Hoegh-Guldberg O, Mumby PJ, Hooten AJ, Steneck RS, Greenfield P, Gomez E, Harvell CD, Sale PF, Edwards AJ, Caldeira K, Knowlton N, Eakin CM, Iglesias-Prieto R, Muthiga N, Bradbury RH, Dubi A, Hatziolos ME (2007) Coral reefs under rapid climate change and ocean acidification. Science 318:1737–1742

Honma Y, Kitami T (1978) Fauna and flora in the waters adjacent to the Sado Marine Biological Station, Niigata University. Ann Rep Sado Mar Biol Stat Niigata Univ 8:7–81

Hori N (1980) Coral reefs in Japan. Kagaku 50:111–122 (in Japanese)

Ikeda E, Iryu Y, Sugihara K, Ohba H, Yamada T (2006) Bathymetry, biota, and sediments on the Hirota reef, Tane-ga-shima; the northernmost coral reef in the Ryukyu Islands. Island Arc 15: 407–419

Inoue S, Kayanne H, Yamamoto S, Kurihara H (2013) Spatial community shift from hard to soft corals in acidified water. Nat Clim Chang 3:683–687

Kan H, Nakashima Y, Ohashi T, Hamanaka N, Okamoto T, Nakai T, Hori N (2005) Drilling Research of a high-latitude coral reef in Mage Island, Satsunan Islands, Japan. Okayama Univ Earth Sci Rep 12:49–58 (in Japanese with English abstract)

Kayanne H, Harii S, Yamano H, Tamura M, Ide Y, Akimoto F (1999) Changes in living coral coverage before and after the 1998 bleaching event on coral reef flats of Ishigaki Island, Ryukyu Islands. Galaxea JCRS 1:73–82 (in Japanese with English abstract)

Kleypas JA, McManus JW, Meñez LAB (1999a) Environmental limits to coral reef development: where do we draw the line? Am Zool 39:146–159

Kleypas JA, Buddemeier RW, Archer D, Gattuso J-P, Langdon C, Opdyke BN (1999b) Geochemical consequences of increased atmospheric carbon dioxide on coral reefs. Science 284: 118–120

Kleypas JA, Feely RA, Fabry VJ, Langdon C, Sabine CL, Robbins LL (2006) Impacts of ocean acidification on coral reefs and other marine calcifiers: a guide for future research, report of a workshop held 18–20 Apr 2005, St. Petersburg, sponsored by NSF, NOAA, and the U.S. Geological Survey, 88 pp

Langdon C, Takahashi T, Sweeney C, Chipman D, Goddard J, Marubini F, Aceves H, Barnett H, Atkinson MJ (2000) Effect of calcium carbonate saturation state on the calcification rate of an experimental coral reef. Glob Biogeochem Cycles 14:639–654

Langdon C, Broecker WS, Hammond DE, Glenn E, Fitzsimmons K, Nelson SG, Peng T-H, Hajdas I, Bonani G (2003) Effect of elevated CO_2 on the community metabolism of an experimental coral reef. Glob Biogeochem Cycles 17:1011

Manzello DP (2010) Coral growth with thermal stress and ocean acidification: lessons from the eastern tropical Pacific. Coral Reefs 29:749–758

Meissner KJ, Lippmann T, Sen Gupta A (2012) Large-scale stress factors affecting coral reefs: open ocean sea surface temperature and surface seawater aragonite saturation over the next 400 years. Coral Reefs 31:309–319

Morita M, Suwa R, Iguchi A, Nakamura M, Shimada K, Sakai K, Suzuki A (2009) Ocean acidification reduces sperm flagellar motility in broadcast spawning reef invertebrate. Zygote 18:103–107

Orr JC, Fabry VJ, Aumont O, Bopp L, Doney SC, Feely RA, Gnanadesikan A, Gruber N, Ishida A, Joss F, Key RM, Lindsay K, Maier-Reimer E, Matear R, Monfray P, Mouchet A, Najjar RG, Plattner GK, Rodgers KB, Sabine CL, Sarmiento JL, Schlitzer R, Slater RD, Totterdell IJ, Weirig MF, Yamanaka Y, Yool A (2005) Anthropogenic ocean acidification over the twenty first century and its impact on calcifying organisms. Nature 437:681–686

Pandolfi JM, Connolly SR, Marshall DJ, Cohen AL (2011) Projecting coral reef futures under global warming and ocean acidification. Science 333:418–422

Precht WF, Aronson RB (2004) Climate flickers and range shifts of reef corals. Front Ecol Environ 2:307–314

Riegl B, Piller WE (2003) Possible refugia for reefs in times of environmental stress. Coral Reefs 92:520–531

Rodgers KB, Aumont O, Menkes C, Gorgues T (2008) Decadal variations in equatorial Pacific ecosystems and ferrocline/pycnocline decoupling. Glob Biogeochem Cycles 22, GB2019

Steinacher M, Joos F, Frölicher TL, Plattner G-K, Doney SC (2009) Imminent ocean acidification in the Arctic projected with the NCAR global coupled carbon cycle-climate model. Biogeosciences 6:515–533

Steinacher M, Joos F, Stocker TF (2013) Allowable carbon emissions lowered by multiple climate targets. Nature 499:197–201

Trenberth KE, Jones PD, Ambenje P, Bojariu R, Easterling D, Klein Tank A, Parker D, Rahimzadeh F, Renwick JA, Rusticucci M, Soden B, Zhai P (2007) Observations: surface and atmospheric climate change. In: Solomon S, Qin D, Manning M, Chen Z, Marquis M, Averyt KB, Tignor M, Miller HL (eds) Climate change 2007: the physical science basis. Contribution of working group I to the fourth assessment report of the intergovernmental panel on climate change. Cambridge University Press, Cambridge

van Hooidonk R, Maynard JA, Planes S (2013) Temporary refugia for coral reefs in a warming world. Nat Clim Chang 3:508–511

Veron JEN, Minchin PR (1992) Correlations between sea surface temperature, circulation patterns and the distribution of hermatypic corals of Japan. Cont Shelf Res 12:835–857

Yamamoto-Kawai M, McLaughlin FA, Carmack EC, Nishino S, Shimada K (2009) Aragonite undersaturation in the Arctic Ocean: effects of ocean acidification and climate change. Science 326:1098–1100

Yamano H, Sugihara K, Nomura K (2011) Rapid poleward range expansion of tropical reef corals in response to rising sea surface temperatures. Geophys Res Lett 38, L04601

Yara Y, Vogt M, Fujii M, Yamano H, Hauri C, Steinacher M, Gruber N, Yamanaka Y (2012) Ocean acidification limits temperature-induced poleward expansion of coral habitats around Japan. Biogeosciences 9:4955–4968

Part II
Advanced Methods of Biodiversity Monitoring

Chapter 5
Classification of Seagrass Beds by Coupling Airborne LiDAR Bathymetry Data and Digital Aerial Photographs

Satoshi Ishiguro, Katsumasa Yamada, Takehisa Yamakita, Hiroya Yamano, Hiroyuki Oguma, and Tsuneo Matsunaga

Abstract Evaluation of the spatial distribution pattern of patchy and fragmental seagrass beds, as hotspots of faunal biodiversity and of high primary productivity, is key to the robust understanding of the ecological state and of the effects of environmental changes on biota in coastal areas. Supervised classification of aerial photographs and satellite imagery is used for assessing the state of shallow-water bottom features (i.e., substrata), such as rock and seagrass patches. For accurate classification, it is important to measure the topography of the seabed extensively and at high resolution, because the color of aerial photographs must be corrected for depth. This is difficult, however, because the shallowness of the water restricts the movements of survey vessels. We generated a digital surface model (DSM) of shallow-water bottom features via airborne LiDAR bathymetry and then used the DSM and digital aerial photographs to classify the bottom features. We conducted simultaneous bathymetry and aerial photography of a bay on the east coast of Tohoku, Japan, using a Fugro LADS Mk 3 system for bathymetry (at 5-m resolution) and a RedLake image sensor for aerial photography (at 0.4-m resolution). After using the topographic data to correct for absorption, we classified the imagery

S. Ishiguro (✉) • H. Yamano • H. Oguma • T. Matsunaga
National Institute for Environmental Studies, Onogawa 16-2, Tsukuba 305-8506, Japan
e-mail: ishiguro.satoshi@nies.go.jp

K. Yamada
Research Center for Fisheries and Environment in the Ariake and Yatsushiro Bay, Seikai National Fisheries Research Institute, Fisheries Research Agency, Taira-machi, Nagasaki 851-2213, Japan

T. Yamakita
Japan Agency for Marine Earth, Science and Technology, Institute of Biogeosciences, Natsushimacho 2-15, Yokosuka 237-0061, Japan

S.-i. Nakano et al. (eds.), *Aquatic Biodiversity Conservation and Ecosystem Services*, Ecological Research Monographs, DOI 10.1007/978-981-10-0780-4_5

to reveal the distribution of seagrass beds. The estimated distribution corresponded with empirical observations.

Keywords Airborne LiDAR • Supervised classification • Shallow-water bottom features • Absorption correction • Seagrass

Introduction

Seagrass generally enhances faunal diversity by increasing habitat complexity, providing living space and shelter for a great variety of animal species. This is because seagrass leaves provide habitats and a wide range of food sources (e.g., epiphytic microalgae, phytoplankton, and seagrass detritus) for a variety of species. As a result, seagrass beds in coastal areas are regarded as hotspots of faunal biodiversity (e.g., Hemminga and Duarte 2000; Larkum et al. 2006; Yamada et al. 2007, 2011; Yamakita and Miyashita 2014). Therefore, the evaluation of the spatial distribution pattern of patchy seagrass beds, as hotspots of faunal biodiversity and of high primary productivity, is key to the robust understanding of the ecological state and of the effect of environmental changes on biota in coastal areas (e.g., Hemminga and Duarte 2000; Waycott et al. 2009). However, mapping the patchy and fragmental distribution of seagrass beds in shallow coastal areas is difficult, because the collection of sea-truth data depends mostly on observation by scuba at discrete points.

Methods for mapping such patchy seagrass by supervised classification using aerial photography and satellite imagery have been proposed (e.g., Wabinaz et al. 2008; Yamakita and Nakaoka 2009; Waycott et al. 2009; Sakuno and Kunii 2013). Seagrass inhabits shallow areas of a few meters' depth. The absorption of light by water, which depends on depth, makes the knowledge of water depth crucial for accurate and homogeneous classification (Sugimori and Sakamoto 1990; Sakamoto et al. 2012). Bathymetry by side-scan sonar can provide depth, but it is not possible in shallow water and among fishing nets or buoys, and it is difficult to obtain aerial photographs simultaneously. In contrast, airborne LiDAR bathymetry (ALB) can measure the depth of shallow water at the same time as digital aerial photographs are taken.

We generated a digital surface model (DSM) of shallow-water bottom features via ALB, corrected the digital aerial photographs for atmospheric scattering and water absorption, classified the imagery by the method of supervised classification, and compared the results of the classification with raw images to assess the accuracy of the classification.

Materials and Methods

Airborne LiDAR Bathymetry System with Digital Image Sensor

ALB systems can measure the depth of water shallower than nominally 2.5× the Secchi disk depth at 532 nm. We used a Fugro LADS Mk 3 ALB system, which acquires data on a 5-m grid (max. resolution 2 m). It has a swath width of ~600 m at a height of ~900 m (Table 5.1). We conducted our survey with a swath width of 330 m. For aerial photography, we used a RedLake image sensor, which has a 0.4-m resolution and obtains imagery in RGB color (Table 5.2). Images are captured at fixed rate of once per second, synchronous with the laser firing. The images are saved in RGB JPEG format with a time-stamped filename. The geo-referenced images are integrated with the Fugro LADS Processing System, allowing the operator to interactively see the location of the laser sounding within the image.

As part of a research project to evaluate the risk of giant earthquakes and tsunami, we conducted simultaneous bathymetry and aerial photography of the Tohoku area in Japan, including Yamada Bay (Fig. 5.1). Additional topographic data came from the M7000-series 1-m-interval isobath data set of the Japan Coast Guard. Figure 5.2 shows an example of LiDAR point data obtained in Yamada Bay.

Because the 3D point data obtained by LiDAR are based on ellipsoidal height, we must subtract the geoid height to relate the altitude to mean sea level. The GSIGEO 2000 data set compiled by the Geospatial Information Authority of Japan provided the geoid data (Kuroishi et al. 2002). The tide level at the time the images were acquired was obtained from the Japan Meteorological Agency's website to convert altitude to water depth (JMA 2013). The tide level was +0.2 m, so the water

Table 5.1 Specifications of Fugro LADS system

Laser rate	1.5 kHz
Depth range	0–80 m
	Nominally 2.5× Secchi disk depth
Topographic range	0–50 m above sea level
Operating altitude	366–914 m
Aircraft speed	232–324 km/h
Scan pattern	Rectilinear
Swath width	Dependent on scan pattern: nominal 71–600 m (independent of aircraft height and water depth)
Relative reflectivity	0–255 value for per-pulse seabed reflectivity at 532 nm; values are relative and scaled across entire survey to maximize dynamic range
Image sensor	RedLake MegaPlus II ES 2020 high-speed digital camera with ultrawide-angle lens (see also Table 5.2)

Table 5.2 RedLake image sensor specifications

Capture rate	1 Hz
Resolution at sea surface	15 cm at 366 m
	20 cm at 488 m
	37 cm at 914 m
	Merged to mosaic with nominally 0.4 m per pixel
Accuracy	5 m 95 % circular error probability
Imaging device	KAI-2020, CCD, color, and mono
Sensor readout	Interline, progressive scan
Resolution	1,600 × 1,200, 1.9 megapixels
Imager size (diagonal)/active area	14.78 mm/11.8 × 8.9 mm

depth was calculated by adding 0.2 m to the altitude. The DSM resolution was interpolated to fit that of the RedLake imagery by cubic convolution.

Study Area and Field Observation

Yamada Bay, on the northeast coast of Japan, is part of a ria coast (Fig. 5.1). There are two islands in Yamada Bay, named Oshima and Koshima. Around these small islands, the seagrass habitat was known from before the Great East Japan Earthquake and tsunami disaster of March 2011. In a series of scuba dives in August 2012, we photographed the bottom features at seven sites, recorded the seagrass species and the coverage of sand and gravel, and logged the latitude and longitude of each location by GPS. We measured the coverage of seagrasses (%) in square quadrats (0.5 × 0.5 m): we placed the quadrats randomly at each GPS point and recorded the coverage (0–100 % in 5 % intervals) of each species with >5 replications. Mean coverage (%) at each GPS point was grouped into four classes: *dense* (>80 % seagrass), *moderate* (30–80 %), *sparse* (10–30 %), and *bare* (<10 %) (Fig. 5.3).

Image Processing

Each pixel value in a RedLake image is given as a digital number (DN) of the RGB value (DN(R,G,B)). It must be corrected for atmospheric scattering and for water absorption. For the former, we used a dark-object subtraction technique (Chavez 1988), in which the darkest pixel in the marine area of an image is subtracted from all pixel values.

For the latter, we assumed that DN is linearly related to irradiance and that DN (R,G,B) is a function of the extinction coefficient and water depth:

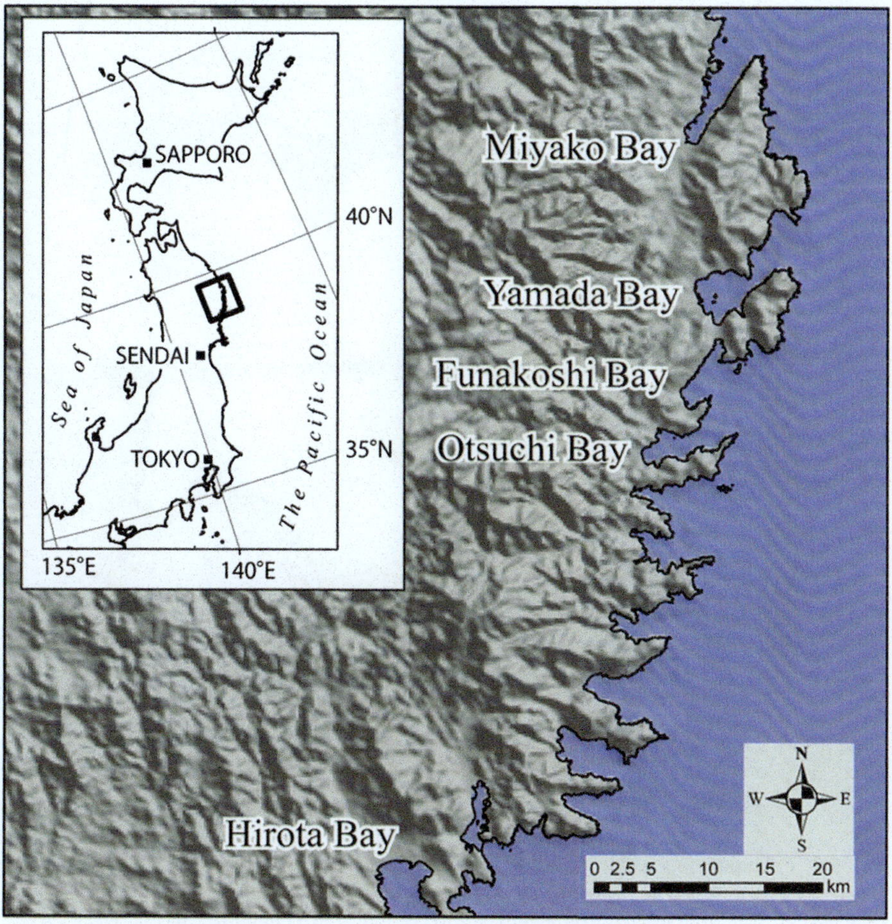

Fig. 5.1 Study area. Yamada Bay is part of the ria coast of northeast Japan

$$\mathrm{DN(R,G,B)} = \mathrm{C} \times \exp[-2 \times k(\mathrm{R,G,B}) \times z] \tag{5.1}$$

where k(R,G,B) is the extinction coefficient in each color band, z is the water depth (m), and C is a constant. To determine the extinction coefficient, we plotted water depth against DN(R,G,B) at sites with similar bottom properties and fitted exponential curves (Fig. 5.4). Values were estimated as $k(\mathrm{R}) = 0.087$, $k(\mathrm{G}) = 0.0885$, and $k(\mathrm{B}) = 0.074$.

The DN values at all pixels were corrected for depth as:

$$\mathrm{DN}_{(\mathrm{R\_rev})} = \frac{\mathrm{DN}_{(\mathrm{R})}}{\exp[-2 \times 0.087 \times \mathrm{z}]} \tag{5.2}$$

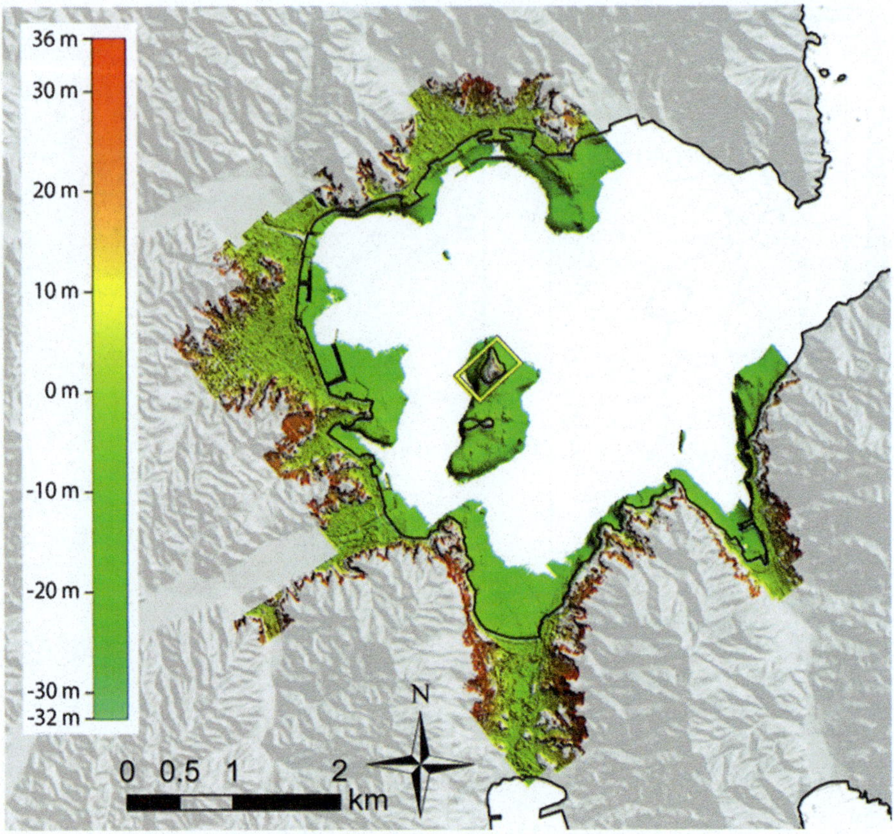

Fig. 5.2 Example of LiDAR point data obtained in Yamada Bay. Topographic data were generated by 10-m digital elevation model by the Geospatial Information Authority of Japan. Data deeper than 30 m are not available. *Yellow rectangle* shows boundary of Fig. 5.5

$$DN_{(G\_rev)} = \frac{DN_{(G)}}{\exp[-2 \times 0.0885 \times z]} \tag{5.3}$$

$$DN_{(B\_rev)} = \frac{DN_{(B)}}{\exp[-2 \times 0.074 \times z]} \tag{5.4}$$

Comparison of original and corrected images shows that the result is uniformly colored, with no dependence of water depth (Fig. 5.5).

Then we conducted supervised classification using a maximum likelihood classification method in ArcGIS software (v. 10.1, ESRI Inc.). The supervised data were set according to sea-truth data collected by scuba and visual observation from a boat (Fig. 5.6). The supervised data comprise two *dense* sites of 1.67 m^2, two *moderate* sites of 1.54 m^2, three *sparse* sites of 3.57 m^2, and one *bare* site of 3.58 m^2. We used

Fig. 5.3 Underwater photographs of sites classified as (**a**) *dense*, (**b**) *moderate*, (**c**) *sparse*, and (**d**) *bare* in Yamada Bay in August 2012. *ZM Zostera marina, PI Phyllospadix iwatensis*

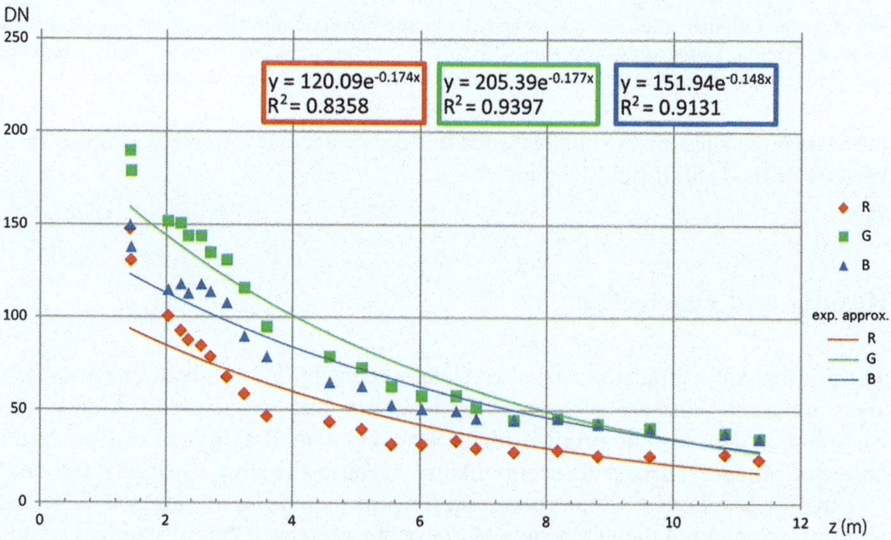

Fig. 5.4 Scatter plots of absorbed radiation and fitted equations

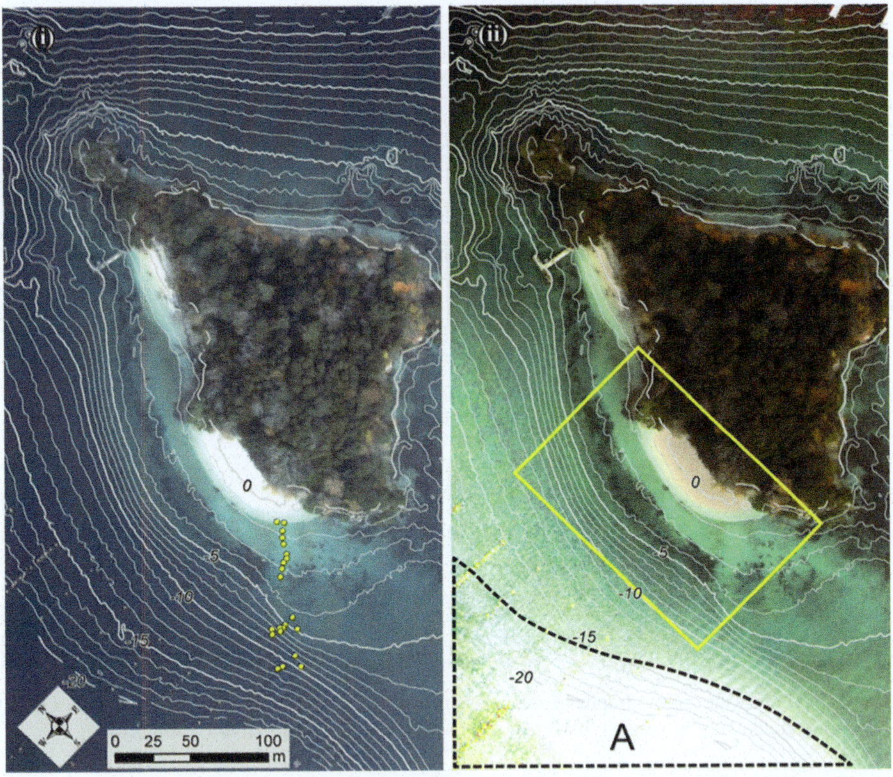

Fig. 5.5 (**a**) Original RedLake image and (**b**) image corrected for atmospheric scattering and water absorption. *Yellow dots* show sites with similar *bottom* properties (possibly sandy area). The *yellow rectangle* marks the analysis area. The deep area marked *A* has been overcorrected

the same eight sites to sea-truth by scuba to validate accuracy, in place of novel sites, because of the lack of rich site data.

Results and Discussion

The original and corrected images were classified (Fig. 5.7). In both images, *dense* areas are patchy and are surrounded by *moderate* and *sparse* areas. A deep area classified as *dense* in the original image was reclassified as *sparse* or *bare* in the corrected image. This result seems likely, because the area is largely too deep (>7 m) for seagrasses. Color correction for atmospheric scattering and for water absorption increased the percentage of *dense* but decreased that of *sparse*. Overall, color correction improved the accuracy slightly from 94.5 to 95.5 % (Table 5.3).

Supervised classification successfully classified areas of seagrass coverage from very-high-resolution bathymetry data and digital aerial photographs. The pattern of

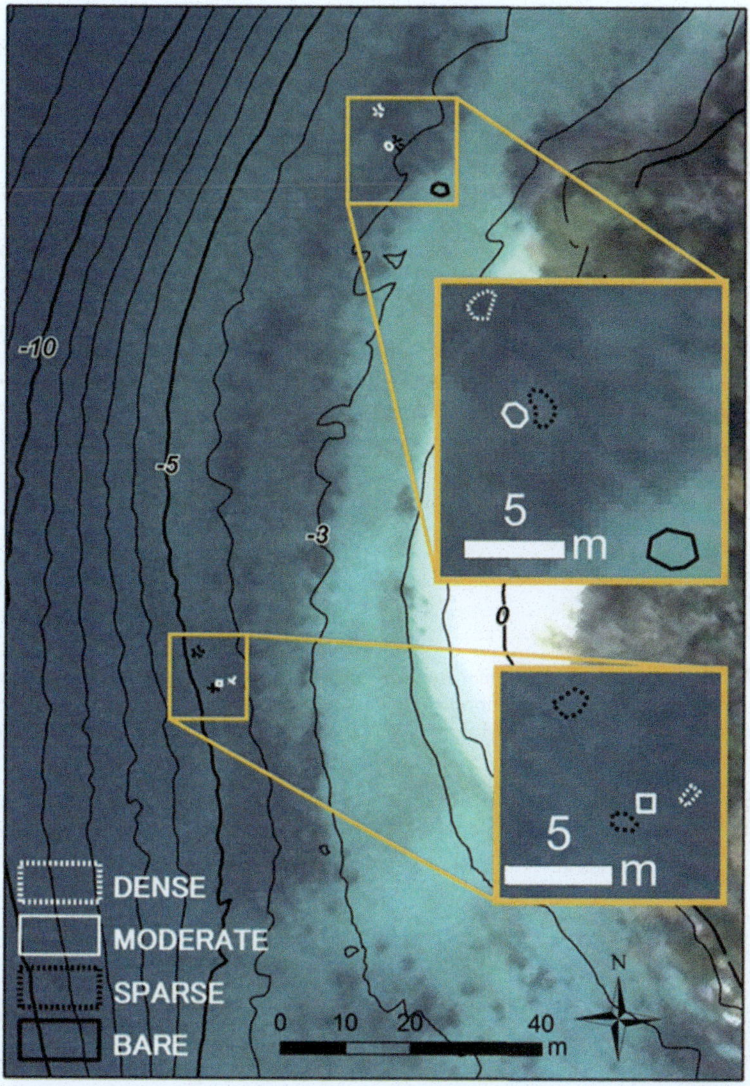

Fig. 5.6 Sites of supervised data (sea-truthed by scuba) in original RedLake image. The contours were generated from ALB data

spatial distribution of each class corresponds to empirical field observations. Our method and data sets will be useful for the quantitative analysis of seagrass distribution and density. However, some block noise was evident ("C" in Fig. 5.7a). Normally, the extinction coefficient of R (red) should be larger than that of G (green). However, the result was the opposite. The RedLake imager has automatic gain control, and its proprietary conversion from irradiance to DN values may use a nonlinear algorithm.

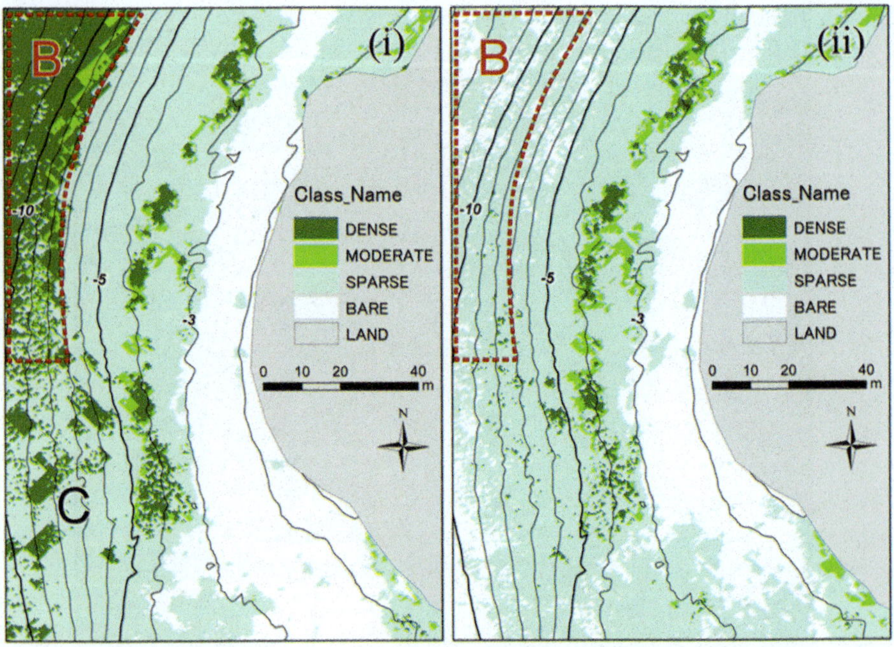

Fig. 5.7 (**a**) Classification of original RedLake image. The deep area marked *B* was classified mostly as *dense*. *C* seems block noise. (**b**) Classification of image corrected for atmospheric scattering and water absorption. The deep area marked *B* was classified as *sparse* or *bare*

Table 5.3 Accuracy before and after color correction of RedLake image. Accuracy values are quotient of actual area divided by estimated area in supervised area. Values show before and after

Actual\estimate	Dense	Moderate	Sparse	Bare	Overall
Dense	0.92, 1.00	0.08, 0.00			
Moderate		0.90, 0.90	0.10, 0.10		
Sparse		0.04, 0.08	0.96, 0.92		
Bare				1.00, 1.00	
Overall					0.945, 0.955

Compared with digital aerial photography without bathymetry calibration, image correction using DSM obviously improved the identification of the seagrass distribution, especially in deeper areas (Fig. 5.7). Because of the darker state of deeper areas, locations without seagrass tended to be identified as seagrass in RGB images.

In general, mapping patchy seagrass distribution is important: it enables us to evaluate biomass and total production (e.g., Green et al. 2000; Duffy 2006; Sagawa et al. 2008), effects of disturbance (e.g., Sleeman et al. 2005; Yamakita et al. 2011), potential distribution and habitat competition among faunal species (e.g., Hovel and Fonseca 2005; Heck and Valentine 2006; Boström et al. 2006; Yamada and

Kumagai 2012; Yamakita and Miyashita 2014), and spatial variation of ecosystem functions (i.e., faunal functional diversity) leading to ecosystem services (Duffy 2006; Yamada et al. 2007, 2011). Therefore, mapping spatial distribution patterns of patchy seagrass beds is key to a robust understanding of the ecological state and the effect of environmental conditions on faunal communities.

Conclusion

We mapped the spatial distribution of seagrass coverage in Yamada Bay by the coupled use of high-resolution digital aerial photography and a topographic-data-based ALB survey. The estimated distribution corresponded with empirical observations. This method will contribute to the monitoring of the spatial dynamics of seagrass patches from shallow to deep areas. Our investigations are continuing in other bays of the Tohoku region.

Acknowledgments We gratefully thank Y. Mochizuki, M. Nakaoka, M. Tamaoki, N. Nakajima, N. Takamura, M. Kurosawa, A. Shirai, and the staff at the International Coastal Research Center of the Ocean Research Institute (University of Tokyo) for the help with field data collection. This research was supported in part by the Center Project of the National Institute for Environmental Studies and Center for Environmental Biology and Ecosystem Studies (no. 1112AF001) and conducted by the National Institute of Advanced Industrial Science and Technology (no. 1112ZZ002).

References

Boström C, Jackson EL, Simenstad CA (2006) Seagrass landscapes and their effects on associated fauna: a review. Estuar Coast Shelf Sci 68:383–403

Chavez PS Jr (1988) An improved dark-object subtraction technique for atmospheric scattering correction of multispectral data. Remote Sens Environ 24:459–479

Duffy EJ (2006) Biodiversity and the functioning of seagrass ecosystems. Mar Ecol Prog Ser 311:233–250

Green EP, Mumby PJ, Edwards AJ, Clark CD (2000) Remote sensing handbook for tropical coastal management. UNESCO, Paris

Heck KL Jr, Valentine JF (2006) Plant–herbivore interactions in seagrass meadows. J Exp Mar Biol Ecol 330:420–436

Hemminga MA, Duarte CM (2000) Seagrass ecology. Cambridge University Press, Cambridge

Hovel KA, Fonseca MS (2005) Influence of seagrass landscape structure on the juvenile blue crab habitat-survival function. Mar Ecol Prog Ser 300:170–191

Kuroishi Y, Ando H, Fukushima Y (2002) A new hybrid geoid model for Japan, GSIGEO2000. J Geodesy 76:428–436

JMA (Japan Meteorological Agency) (2013) Tidal data web site (in Japanese). Available from URL: http://www.data.kishou.go.jp/db/tide/genbo/index.php. Accessed 7 Nov 2013

Larkum AW, Orth RRJ, Duarte CM (eds) (2006) Seagrasses: biology, ecology, and conservation. Springer, Dordrecht

Sagawa T, Mikami A, Komatsu T, Kosaka N, Kosako A, Miyazaki S, Takahashi M (2008) Mapping seagrass beds using IKONOS satellite image and side scan sonar measurements: a Japanese case study. Int J Remote Sens 28:281–291

Sakamoto SX, Sasa S, Sawayama S, Tsujimoto R, Terauchi G, Yagi H, Komatsu T (2012) Impact of huge tsunami in March 2011 on seaweed bed distributions in Shizugawa Bay, Sanriku Coast, revealed by remote sensing. Proc SPIE 8525. doi:10.1117/12.999308

Sakuno Y, Kunii H (2013) Estimation of growth area of aquatic macrophytes expanding spontaneously in Lake Shinji using ASTER data. Int J Geosci 4:1–5

Sleeman JC, Kendrick GA, Boggs GS, Hegge BJ (2005) Measuring fragmentation of seagrass landscapes: which indices are most appropriate for detecting change? Mar Freshw Res 56:851–864

Sugimori Y, Sakamoto W (1990) Ocean environmental optical science. Tokai University Press, Tokyo (in Japanese)

Wabnitz CC, Andréfouët S, Torres-Pulliza D, Müller-Karger FE, Kramer PA (2008) Regional-scale seagrass habitat mapping in the Wider Caribbean region using Landsat sensors: applications to conservation and ecology. Remote Sens Environ 112:3455–3467

Waycott M, Duarteb CM, Carruthers TJB, Orth RJ, Dennison WC, Olyarnick S, Calladine A, Fourqurean JW, Heck KL Jr, Randall Hughes A, Kendrick GA, Judson Kenworthy W, Short FT, Williams SL (2009) Accelerating loss of seagrasses across the globe threatens coastal ecosystems. Proc Natl Acad Sci U S A 106:12377–12381

Yamada K, Kumagai NH (2012) Importance of seagrass vegetation for habitat partitioning between closely related species, mobile macrofauna *Neomysis* (Misidacea). Hydorobiologia 680:125–133

Yamada K, Hori M, Tanaka Y, Hasegawa N, Nakaoka M (2007) Temporal and spatial macrofaunal community changes along a salinity gradient in seagrass meadows of Akkeshi-ko estuary and Akkeshi Bay, northern Japan. Hydrobiologia 592:345–358

Yamada K, Hori M, Nakaoka M, Hamaguchi M (2011) Temporal and spatial variation of functional-trait composition (functional diversity) of macro-crustacean community in seagrass meadow. Crustaceana Monogr 15:325–339

Yamakita T, Miyashita T (2014) Landscape mosaicness in the ocean: its significance for biodiversity patterns in benthic organisms and fish. In: Nakano S, Yahara T, Nakashizuka T (eds) Integrative observations and assessments (Ecological research monographs/Asia-Pacific biodiversity observation network). Springer Japan, pp 131–148

Yamakita T, Nakaoka M (2009) Scale dependency in seagrass dynamics: how does the neighboring effect vary with grain of observation? Popul Ecol 51:33–40

Yamakita T, Watanabe K, Nakaoka M (2011) Asynchronous local dynamics contributes to stability of a seagrass bed in Tokyo Bay. Ecography 34:519–528

Yamakita T, Taki H, Okabe K, Quantitative effects of terrestrial and oceanic factors on the nationwide distribution of seagrass and Sargasso beds at two different river basin scales (under review)

Chapter 6
Cyanobacterial Blooms as an Indicator of Environmental Degradation in Waters and Their Monitoring Using Satellite Remote Sensing

Yoichi Oyama, Bunkei Matsushita, and Takehiko Fukushima

Abstract Cyanobacterial bloom is a major problem in many lakes and rivers around the world even now in the twenty-first century. It has detrimental effects not only on the aquatic environment such as reduced transparency, elevated pH, and oxygen depletion but also on the drinking, agricultural, industrial, commercial, and recreational uses of inland waters. In this chapter, we first review the influence of the harmful cyanobacterial blooms on aquatic fauna and flora such as zooplankton, fish, and aquatic macrophyte. And then we introduce the monitoring of the cyanobacterial bloom using satellite remote sensing. Satellite remote sensing could present a valuable tool to obtain more reliable information about the extent of the cyanobacterial bloom than the conventional monitoring methods such as ship survey. With the rapid development of satellite sensors, many useful algorisms have been proposed by scientists. As one of the methods, we introduce a novel method for monitoring the abundance of the cyanobacterial bloom from Landsat images using an environmental indicator, namely, the visual cyanobacteria index (VCI).

Keywords Biodiversity • Cyanobacterial bloom • Visual cyanobacteria index (VCI) • Chl-a • Fish • Plankton • Macrophyte • Remote sensing • Landsat • Lake • Water quality • Eutrophication

---

Y. Oyama (✉)
Marimo Research Center, Kushiro Board of Education, Hokkaido, Japan
e-mail: oyama401@gmail.com

B. Matsushita • T. Fukushima
Faculty of Life and Environmental Sciences, University of Tsukuba, Tennoudai 1-1-1, Tsukuba, Ibaraki 305-8572, Japan

S.-i. Nakano et al. (eds.), *Aquatic Biodiversity Conservation and Ecosystem Services*, Ecological Research Monographs, DOI 10.1007/978-981-10-0780-4_6

Introduction

During the last several centuries, since the industrial revolution, inland and coastal waters have been exposed to numerous anthropogenic stress factors. Chemical pollutants increased in number and concentration caused by rising population densities, farming, and industrialization (Hupfer and Hilt 2008). In particular, eutrophication, caused by the excessive inputs of the nutrients (i.e., phosphorus and nitrogen), is recognized as a common and growing problem in the world. In the late twentieth century, 54 % of lakes in Asia Pacific region were eutrophic; the proportions for Europe, Africa, North America, and South America were 53 %, 28 %, 48 %, and 41 %, respectively (ILEC/Lake Biwa Research Institute 1988–1993). The water with higher trophic levels results in the dominance of large, colony-forming species of cyanobacteria (blue-green algae) such as *Microcystis*, *Anabaena*, and *Aphanizomenon* (Dokulil and Teubner 2000). Therefore, permanent cyanobacterial dominance is regarded as the ultimate phase of eutrophication (Robarts 1985; Jones 1994; Pizzolon et al. 1999). Massive growth of cyanobacteria leads to the production of blooms, scums, and mats and has a wide range of negative effects not only on aquatic ecosystems (e.g., species diversity) but on human activities (e.g., water resources).

In the twenty-first century, a movement has started to assess the occurrence of the cyanobacterial blooms on a global scale. The CYANONET is one of the few database programs to create a global source of information on cyanobacterial mass population, cyanotoxins, and associated human and animal health problems (Codd et al. 2005). In addition, The United Nations Environmental Programme (UNEP) has developed a global database, namely, CyanoData, in order to centralize information about cyanobacterial bloom management. Recently, the satellite remote sensing technique has been used to obtain the information of the extent of cyanobacterial bloom. Gómez et al. (2011) reported a potential use of this technique to the temporal series to monitor the European Union Water Framework Directive (WFD).

In this chapter, we first review the impact of cyanobacterial bloom on aquatic biodiversity and human activities and then introduce the current method for monitoring cyanobacterial bloom using satellite remote sensing.

Impact of Cyanobacterial Bloom on Aquatic Biodiversity

The transition of trophic state from oligotrophic to eutrophic can increase the biological productivity such as photosynthesis by autotrophs and the grazing by consumer species. However, many scientists have pointed out that the eutrophication works negatively on aquatic biodiversity (e.g., Margalef 1968; Seehausen et al. 1997; Abrentes et al. 2006; Smith and Schindler 2009). In actual, the dominance of cyanobacteria that resulted in eutrophication can restrict the growth of other phytoplankton species such as diatoms, chlorophyte, and dinophyte and cause the reduction in phytoplankton diversity. In addition, cyanobacterial blooms have a great impact on the aquatic environment such as reduced transparency, elevated pH, and oxygen depletion (Carpenter et al. 1998; Havens 2007). Moreover, some of cyanobacterial species produce potent toxins (Codd 2000). These environmental alterations may influence other aquatic communities and structures such as zooplankton, fishes, and macrophyte.

Zooplankton Diversity

Zooplankton communities are affected by the composition and abundances of phytoplankton species because they play a role in aquatic food web as primary consumers. In particular, bloom-forming cyanobacteria are low-quality food for zooplankton, due to their large filamentous or colonial structure (DeMott 1999; Abrantes et al. 2006; Havens 2007). Jeppesen et al. (2000) investigated the changes in species richness along a total phosphorus (TP) gradient in 71 Danish lakes. They reported that phytoplankton communities were changed from dinophyte, diatoms, and chrysophytes to cyanophyte with increasing TP while the species richness of all selected zooplankton taxa decreased monotonically. In addition, cyanobacterial toxin can inhibit the growth of zooplankton. *Daphnia* spp. are the most affected species by cyanobacterial bloom because of their high sensitivity to cyanobacterial toxin (Fulton 1988). The cyanobacterial blooms also influence the functional diversity of zooplankton (e.g., trait of adult body size and trophic group). Large zooplankton species (e.g., *Daphnia* spp.) are frequently replaced by small ones (e.g., *Ceriodaphnia* spp.) when cyanobacteria are dominant (Lampert 1981; Barnett and Beisner 2007). This is because the small-bodied zooplankton species can efficiently capture and consume bacteria in the presence of inedible algae (DeMott 1989).

Fish Diversity

When the cyanobacteria blooms decay, they will cause oxygen depletion owing to the microbial decomposition of cyanobacterial cells. The anoxic condition can preclude fish or other biota (e.g., macroinvertebrate) from hypolimnion and bottom sediments, bringing about change in the taxonomic structure of them. Nürnberg (1995) indicates that fish species richness tends to be higher in oxygen saturation condition compared with oxygen depletion condition. In temperate and boreal regions, where piscivores (salmonids) required a cold water refuse during summer, the anoxic condition may eliminate those fishes (Colby et al. 1972). Frey (1955) also reported the loss of cold-water cisco population from Lake Mendota, USA, in the 1930s to 1940s due to hypolimnetic oxygen depletion. Additionally, the availability of benthic macroinvertebrate is also important to the fish communities. Ludsin et al. (2001) investigated the dynamics of fish communities in Lake Erie, USA-Canada, during oligotrophication in 1969–1996. They reported that the recovery of macroinvertebrate prey with declined bottom anoxia may increase the eutrophy-intolerant fish species. Because most recovering intolerant species are cool- or cold-water stenothermal, the enhanced oxygen level in deep, cold bottom waters likely played a pivotal role in their success by enhancing preferred thermal habitat and benthic macroinvertebrate prey.

Aquatic Macrophyte Diversity

Freshwater macrophytes are traditionally classified into four groups based on life form: emergent plants, floating-leaved plants, submerged plants, and free-floating plants (Sculthorpe 1967). The influences of the eutrophication on the macrophytes are different among the groups. In particular, decline and disappearance of submerged plants occurs in many shallow lakes, due mainly to light limitation caused by algal bloom (Scheffer et al. 1993). Geurts et al. (2008) investigated the distribution of macrophytes in 145 fen waters in the Netherlands, focusing on the occurrence of endangered species, and found that macrophytes are restricted to turbidity levels > 40 ppm Pt, whereas the endangered species were almost absent at sites with turbidity levels > 12 ppm Pt. At highly turbid waters, submerged macrophytes will be dominated by canopy-forming species, which will be able to survive under poor light conditions by establishing their biomass just below the water surface (Søndergaard et al. 2013).

The species diversity is also changed with the transition from plant to phytoplankton dominance (Mason and Bryant 1975; Jeppesen et al. 2000; Sayer et al. 2010). Sayer et al. (2010) studied a long-term change in macrophyte and phytoplankton communities in 39 shallow lakes and found a significant negative relationship between chlorophyll-a (Chl-a) concentrations and macrophyte species richness. The plant communities were shifted from a species-rich plant communities with charophytes to a species-poor plant communities dominated by *Potamogeton* spp. These compositional changes are accompanied by a substantial reduction in the seasonal duration of macrophyte dominance and a greater tendency for incursions by phytoplankton. The eutrophic species such as *Potamogeton* spp. are capable of compressing their life cycle into a short early summer phase as stresses associated with abundant phytoplankton populations start to mount (Scheffer and Jeppesen 1998).

Impact of Cyanobacterial Bloom on Human Activities

Cyanobacterial bloom generally has detrimental effects on the drinking, agricultural, industrial, commercial, and recreational uses of inland waters. For example, increased turbidity and particulate matters result in the blockage of water filters and the production of odor compounds in drinking water (Klapper 1991). Unfavorable appearance or unpleasant odor becomes a nuisance to the water user of recreational activities such as boating and swimming (Dodds et al. 2009). Lack of oxygen due to decomposition of organisms causes fish kill and damage to commercial fishery (Dawson 2002).

In particular, cyanobacterial toxins (cyanotoxins) become widely recognized as a human health problem arising as a consequence of eutrophication (Bartram et al. 1999). Contact or ingestion of cyanobacterial cell and toxin can cause skin irritation, gastroenteritis, and liver and kidney damages (Codd 2000). Yu (1995) suggested through long-term studies in China that drinking of untreated surface water containing cyanotoxins is associated with an increased incidence of primary liver cancer. Additionally, the cyanotoxins have caused numerous fatal poisonings of livestock and wildlife. The first report of cyanobacterial poisoning was of the deaths of cattle, sheep, dogs, horses, and pigs after drinking a scum of *Nodularia spumigena* in Lake Alexandrina in Australia (Francis 1878). The Australian government has developed livestock drinking water guidelines for cyanobacteria, providing individual safety standard values (e.g., particular cell counts of *Microcystis* and/or concentrations of microcystins) for the mammals (ANZECC/ARMCANZ 2000).

Recently, there has been an increase in research activities for estimating the environmental costs by assigning economic value to a freshwater ecosystem function or service (e.g., Wilson and Carpenter 1999; Pretty et al. 2003; Dodds et al. 2009). Pretty et al. (2003) estimated the losses of economic values due to the eutrophication using data sets on cyanobacterial bloom in eight water regions of England and Wales. They reported that the annual losses were \$0.7–1.4 million year^{-1} for reduced value of water bodies for commercial use (e.g., livestock watering and irrigation), \$26.6 million year^{-1} for drinking water treatment to remove algal toxin and algal decomposition products, and \$13.51–46.96 million year^{-1} for reduced recreational and amenity value of water bodies. In the case of US freshwaters, the potential annual value losses were \$0.3–2.8 billion year^{-1} for lakefront property values and \$0.37–1.16 billion year^{-1} for recreational uses (Dodds et al. 2009).

Monitoring Cyanobacterial Bloom by Satellite Remote Sensing

Water quality monitoring is usually conducted by the combination of ship survey and laboratory measurement. However it is very difficult to observe the cyanobacterial bloom from traditional ship surveys because of their patchiness and high spatial and temporal variability. Even if the bloom can be found from the ship, the natural spatial distribution of the bloom will be easily destroyed by the ship and sampling devices without special precautions (Kutser 2004). Since the early 1970s, satellite remote sensing has been widely used to monitor the cyanobacterial bloom. Satellite remote sensing constitutes a valuable tool to obtain more reliable information about the extent of the cyanobacterial bloom than the conventional monitoring methods because it makes possible extensive, repetitive, and contactless surveys.

Table 6.1 summarizes the typical remote sensing algorithms and satellite sensors for detecting cyanobacterial blooms. The Baltic Sea is one of the most famous study sites for monitoring cyanobacterial blooms. Karhu et al. (1994) investigated the dynamics of cyanobacterial blooms (dominated by *Nodularia spumigena*) from 1982 to 1993 using Advanced Very High-Resolution Radiometer (AVHRR) images. They detected the bloom using the threshold of reflectance value in the visible AVHRR band (570–700 nm of wavelengths) and reported that more than 20,000 km^2 of water areas were covered by the blooms. Along with the progress

Table 6.1 Examples of remote sensing algorithms for monitoring cyanobacterial blooms

Algorithm	Area	Sensor	References
Single band	Baltic Sea	AVHRR	Karhu et al. (1994)
	Baltic Sea	CZCS	Karhu et al. (2007)
	Baltic Sea	SeaWiFS	Karhu et al. (2007)
	Baltic Sea	MODIS	Karhu et al. (2007)
Empirical or semi-analytical model	South Atlantic Bight	SeaWiFS	Subramaniam et al. (2002)
	Lake Erie	TM	Vincent et al. (2004)
	Baltic Sea	Hyperion	Kutser (2004)
	Spanish and Netherlands Lakes	MERIS	Simis et al. (2005)
	Lac des Allemands	OCM	Dash et al. (2011)
	Lake Omodeo	MERIS	Bresciani et al. (2012)
Vegetation index	Lake Taihu	MODIS	Chen and Dai (2008)
	Yellow Sea, Bohai Sea and East China Sea	MODIS	Hu and He (2008)
Spectral shape	Coast of Vancouver Island	MERIS	Gower et al. (2005)
	Great Lakes	MERIS	Winne et al. (2008)
	Yellow Sea	MODIS	Hu (2009)
	Lake Taihu, Lake Victoria	MERIS	Matthews et al. (2012)

of ocean color satellite sensors such as Coastal Zone Color Sensor (CZCS), Sea-Viewing Wide Field-of-View Sensor (SeaWiFS), and Moderate-Resolution Imaging Spectroradiometer (MODIS), several empirical and semi-analytical algorithms have been developed to retrieve the Chl-a concentration in water bodies and also have been used to monitor the cyanobacterial bloom (e.g., Subramaniam et al. 2002; Kutser 2004). Recently, the pigment phycocyanin (PC), which is a specific photosynthetic pigment of cyanobacteria, is being attempted to detect the cyanobacterial bloom (e.g., Simis et al. 2005; Dash et al. 2011). The Medium-Resolution Imaging Spectrometer (MERIS) is suitable to retrieve the PC concentrations because this sensor can detect radiance/reflectance signal at 620 nm which corresponds to the absorption peak of the PC. One shortcoming of the empirical and semi-analytical Chl-a or PC algorithms is that they rely on the accurate atmospheric corrections. The most atmospheric correction algorithms designed for open ocean assume negligible water reflectance in near-infrared (NIR) region. In case that the aggregation of cyanobacteria forms dense mats and scums, the reflectance at NIR region is not negligible, causing algorithm failure.

In contrast, some vegetation indices such as normalized difference vegetation index (NDVI) and enhanced vegetation index (EVI) can be used to detect the cyanobacterial scum because spectral shapes of the scum are more similar to those of terrestrial vegetation than those of water bodies (Kutser 2004; Hu and He 2008). Additionally, the methods that used spectral peaks at the NIR region such as maximum chlorophyll index (MCI) (Gower et al. 2005), cyanobacteria index (CI) (Wynne et al. 2008), and floating algae index (FAI) (Hu 2009) are less sensitive to the atmospheric corrections compared with Chl-a and PC algorithms. Unfortunately, there is few study to relate the values of these indices with biomass of cyanobacterial scum such as Chl-a or PC concentrations. One reason is the difficulty for measuring the biomass of cyanobacterial scum because they are free-floating on the water surface (Kutser 2004). Oyama et al. (2015) applied a new environmental indicator of cyanobacterial blooms, namely, the visual cyanobacteria index (VCI) (Aizaki et al. 1995), for monitoring the abundance of the blooms from Landsat images. The detail of the VCI and the case study in Lake Kasumigaura, Japan, will be described in the following sections.

Visual Cyanobacteria Index (VCI)

The visual cyanobacteria index (VCI) proposed by Aizaki et al. (1995) is an index for monitoring bloom levels by visual interpretation such as patches, scums, and mats. Figure 6.1 shows photographs of various levels of cyanobacterial blooms. The VCI classifies cyanobacterial blooms into the six levels of aggregation described below:

Fig. 6.1 Photographs of various levels of cyanobacterial blooms in Lake Kasumigaura, Japan. The blooms were classified into six levels based on visual appearance: (**a**) level 1; (**b**) level 2; (**c**) level 3; (**d**) level 4; (**e**) level 5; (**f**) level 6 (All the photographs were from Aizaki et al. (1995a))

Level 1: Blooms cannot be observed visually without tools such as a phytoplankton net or a white container.
Level 2: The bloom spreads slightly as filaments and can be observed visually.
Level 3: The bloom spreads on the water surface and forms patches and scums.
Level 4: The bloom covers the water surface as thin film.
Level 5: The bloom covers the water surface as thick mat.
Level 6: The cyanobacterial mat is crusted and releases a strong odor; at this level it is called a hyperscum (Zohary and Breen 1989). The color can often change from green to white, purple, or blue.

The VCI does not require any instruments and laboratory measurements, enabling quick and easy observations of the cyanobacterial bloom. The Japan's local governments have used the VCI in order to manage the quality of inland waters. In addition, the respective VCI levels are related to the abundance of cyanobacteria. Figure 6.2 shows the relationships between the VCI and Chl-a or PC concentrations in the bloom of *Microcystis aeruginosa* (Oyama et al. 2015). The pigment concentrations increase with the levels of VCI and reach more than 10,000 $\mu g\ l^{-1}$ of Chl-a and PC concentrations at level 6.

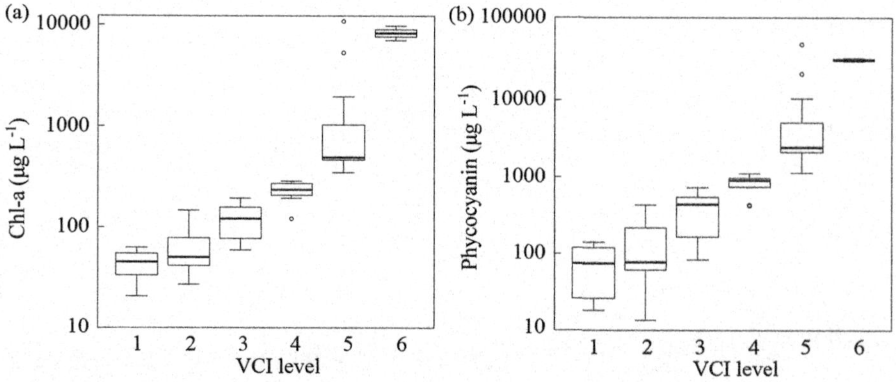

Fig. 6.2 The box-and-whisker plots of pigment concentrations in each VCI level (From Oyama et al. 2015). The vertical axis is expressed by log scale. *Solid lines* in the *boxes* represent the median value. The *upper* and *lower* fences represent the 1st and 3rd quartiles (Q1 and Q3), respectively. The lower and upper whiskers were calculated from (Q1–1.5 × IQR) and (Q3 + 1.5 × IQR), respectively, where IQR is the interquartile range represented by the width of the box (i.e., Q3–Q1). Data above or below the whisker were defined as the outliers and are shown as *open circles*. (**a**) Chl-a and (**b**) phycocyanin

In addition, the reflectance spectra are also varied at the respective VCI levels. Figure 6.3 shows the change in the respective ETM+ band reflectances (except for TM band 6) with VCI levels (Oyama et al. 2015). The ETM+ band reflectances are calculated form in situ-measured reflectance spectra at 1-nm interval. In the case of levels 1 and 2, the spectral peak can be found at band 2 which is corresponding to the green wavelength (Fig. 6.3a, b). The peak is shifted from band 2 to band 4 for VCI level 3 (Fig. 6.3c), and the peak height at band 4 becomes higher with the VCI level (Fig. 6.3d–f). The significant increases at band 5 and 7 are also found from VCI level 4. In particular, the band 5 is higher than band 3 for levels 4 and 5. For VCI level 6, at which the cyanobacteria form a hyperscum, the reflectance at band 4 decreased, whereas it increased at the visible bands (bands 1–3) (Fig. 6.3f). It is due mainly to the decay and drying of the surface scum.

Application of the VCI to Landsat Image in Lake Kasumigaura

Since the reflectance at NIR band (i.e., ETM+ band 4) is sensitive to the VCI level, it can be used to classify the abundance of cyanobacterial blooms into the respective VCI levels from satellite data. The FAI (Hu 2009) is suitable to estimate the magnitude of reflectance at band 4 because it is calculated based on the peak height at band 4 from a baseline between neighboring bands (e.g., TM bands 3 and 5) as follows:

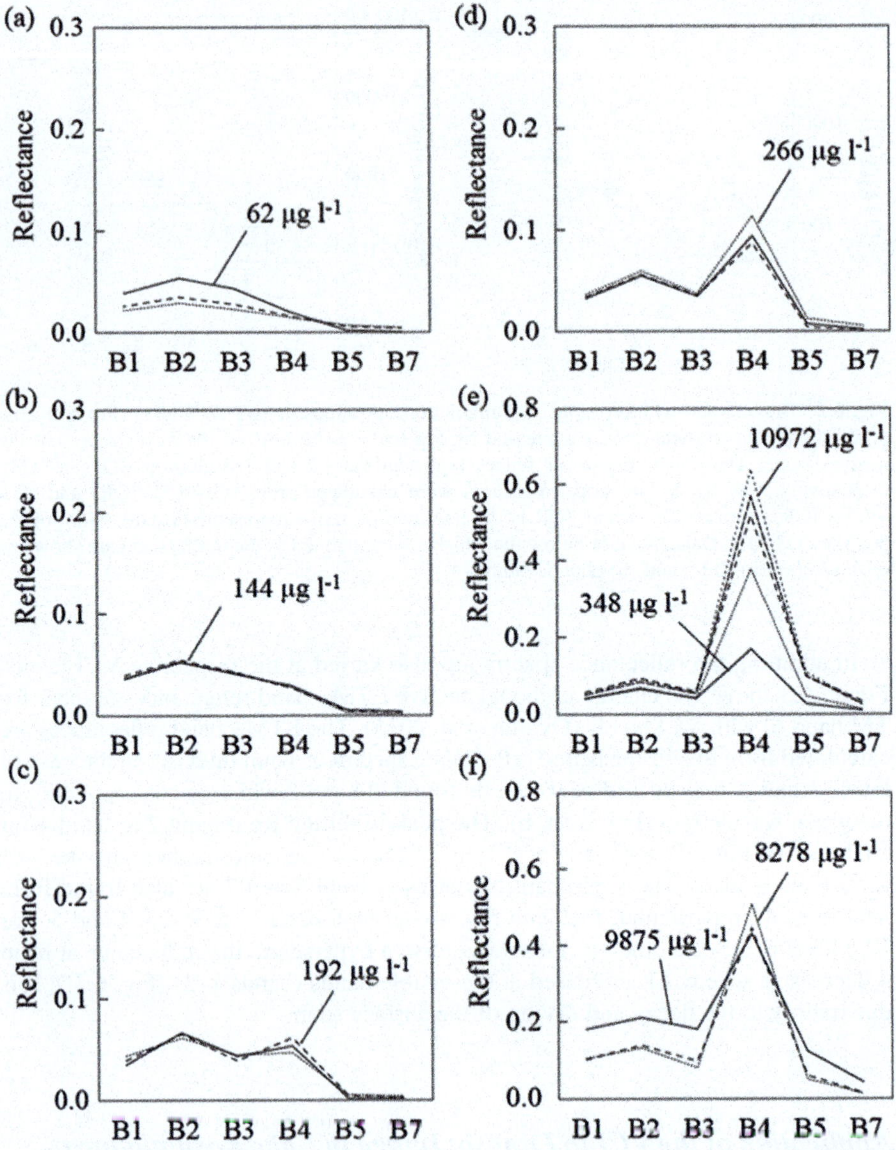

Fig. 6.3 Landsat/TM band reflectances for each level of the VCI level (From Oyama et al. 2015). The numerical values in the figures represent Chl-a concentrations. (**a**) VCI level 1, (**b**) level 2, (**c**) level 3. (**d**) level 4, (**e**) level 5, and (**f**) level 6. The y axis in (**d–f**) is four times as high as that in (**a–c**)

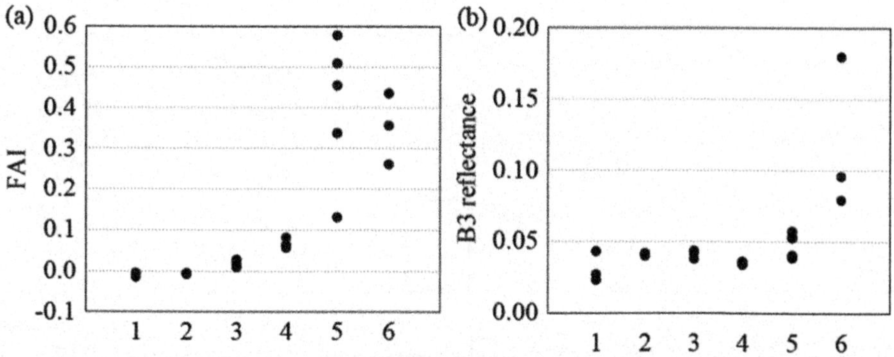

Fig. 6.4 Changes in the FAI and reflectance values at each VCI level. (**a**) FAI and (**b**) reflectance at band 3

$$\mathrm{FAI} = R_{\mathrm{B4}} - \left[R_{\mathrm{B3}} + (R_{\mathrm{B5}} - R_{\mathrm{B3}}) \times \frac{(\lambda_{\mathrm{B4}} - \lambda_{\mathrm{B3}})}{(\lambda_{\mathrm{B5}} - \lambda_{\mathrm{B3}})} \right] \quad (6.1)$$

where R is the Rayleigh-corrected reflectance defined by Hu et al. (2004) and $\lambda_{\mathrm{B}i}$ is the center wavelength for the ith band (i.e., $\lambda_{\mathrm{B3}} = 660$ nm, $\lambda_{\mathrm{B4}} = 825$ nm, and $\lambda_{\mathrm{B5}} = 1650$ nm for ETM+).

Figure 6.4 shows the changes in the FAI and ETM+ band 3 reflectance with VCI levels (Oyama et al. 2015). The FAI can classify the VCI levels into four groups (levels 1 and 2, level 3, level 4, and levels 5 and 6), although it cannot separate between VCI levels 5 and 6 (Fig. 6.4a). Alternatively, reflectance at band 3 differed significantly from that at VCI levels 5 and 6 (Fig. 6.4b). Accordingly, a threshold can be determined between VCI levels by the above result. Figure 6.5 shows the Landsat/TM image and the distribution map of the VCI in west bay of Lake Kasumigaura, Japan, on 11 August 2011. The massive cyanobacterial bloom occurred in the lake for the first time during the past 13 years. The satellite image is shown as a false color (band 3, blue; band 4, green; band 5, red) to enhance the cyanobacterial bloom in the image. However, it is difficult to obtain the information on the degree of cyanobacterial bloom in this image. The distribution map for VCI developed by FAI and TM band 3 reflectance showed that about 4 km^2 of surface water is covered by surface cyanobacterial blooms (more than VCI level 3) (Fig. 6.5b). The VCI level 6, indicating the ultimate phase of cyanobacterial bloom, is sparsely distributed along with the coast. Since the unpleasant odor from decaying cyanobacteria is the serious problem for residents near the water-front, the VCI image is useful to determine the area where any countermeasures (e.g. removal of cyanobacteria) are required.

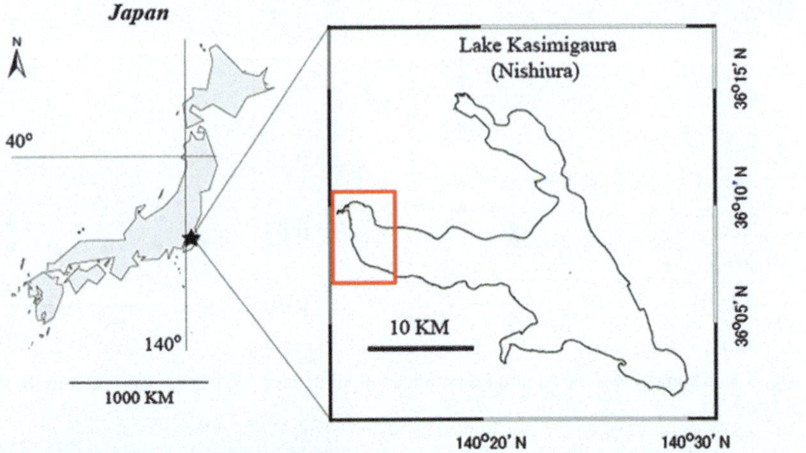

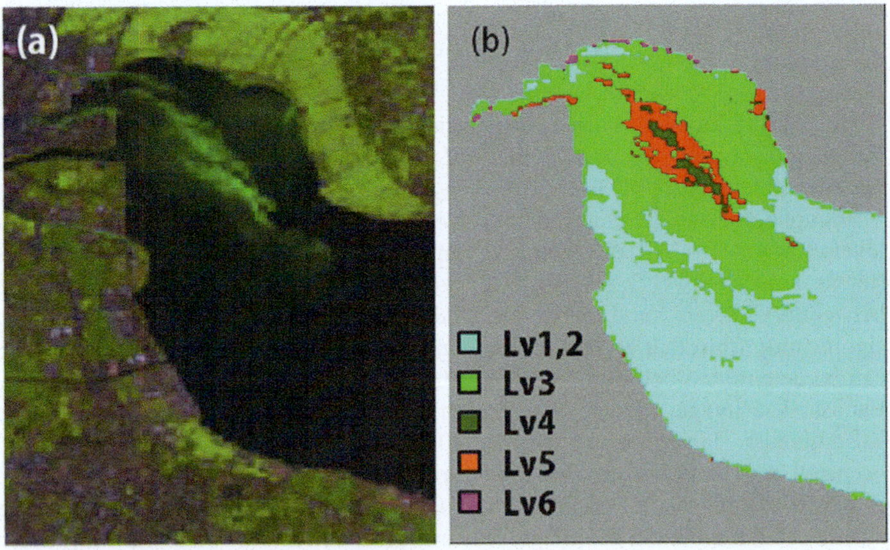

Fig. 6.5 Landsat/TM image in Lake Kasumigaura, Japan, in August 2011. (**a**) False color image (band 3, *blue*; band 4, *green*; band 5, *red*) and (**b**) VCI image

Conclusions

Cyanobacterial bloom is a major problem in many lakes and rivers around the world even now in the twenty-first century. It has negative impacts not only on communities and structures of aquatic flora and fauna but also on human activities such as drinking, agricultural, industrial, commercial, and recreational uses of inland waters. There is an urgent need to construct observation systems in order to protect the freshwater ecosystems and resources. Satellite remote sensing is one of the appropriate tools to accomplish the purpose. It can provide the information on spatial and temporal variability of the cyanobacterial bloom without any field observations. This advantage will help us to find out the water area where biodiversity and ecosystem service are deteriorated.

Acknowledgments The present study was supported by the Environment Research and Technology Development Fund (S9-4-1) of the Ministry of the Environment, Japan.

References

Abrantes N, Antunes MJ, Pereira MJ, Goncalves F (2006) Seasonal succession of cladocerans and phytoplankton and their interactions in a shallow eutrophic lake (Lake Vela, Portugal). Acta Oecol 29(1):54–64

Aizaki M, Fukushima T, Takagi H, Kitamura H (1995) Evaluation of Lake Kasumigaura, Japan, using a landscape index for cyanobacterial bloom. In: Aizaki M, Fukushima T (eds) Aoko (water-blooms of blue-green algae), Measurement, Occurrence, and Factors on Its Growth. National Institute for Environmental Studies, Tsukuba, pp 33–39 (in Japanese)

ANZECC and ARMCANZ (2000) Australian and New Zealand guidelines for fresh and marine water quality. National water quality management strategy paper No 4. Australian and New Zealand Environment and Conservation Council and Agriculture and Resource Management Council of Australia and New Zealand, Canberra

Barnett A, Beisner BE (2007) Zooplankton biodiversity and lake trophic state: explanations invoking resources abundance and distribution. Ecology 88(7):1675–1686

Bartram J, Carmichael WW, Chorus I, Jones G, Skulberg OM (1999) Introduction. In: Chorus I, Bartram J (eds) Toxic Cyanobacteria in water: a guide to their public health consequences, monitoring and management. E & FN Spon, London, pp 1–14

Bresciani M, Vascellari M, Giardino C, Matta E (2012) Remote sensing supports the definition of the water quality status of Lake Omodeo. Eur J Remote Sens 45:349–360

Carpenter SR, Caraco NF, Correll DL, Howarth RW, Sharpley AN, Smith VH (1998) Nonpoint pollution of surface waters with phosphorus and nitrogen. Ecol Appl 8(3):559–568

Chen Y, Dai J (2008) Extraction methods of cyanobacteria bloom in Lake Taihu based on RS data. J Lake Sci 20:179–183 (in Chinese with English abstract)

Codd GA (2000) Cyanobacterial toxins, the perception of water quality, and the prioritization of eutrophication control. Ecol Eng 16(1):51–60

Codd GA, Azevedo SMFO, Bagchi SN, Burch MD, Carmichael WW, Harding WR, Kaya K, Utkilen HC (2005) CYANONET: a global network for cyanobacterial bloom and toxin risk management. International Hydrological Programme: Initial Situation Assessment and Recommendations. IHP-VI technical documents in hydrobiology 76. UNESCO, Paris, pp 1–138

Colby PJ, Spangler GR, Hurley DA, McCombie AM (1972) Effect of eutrophication on salmonid communities in oligotrophic lakes. J Fish Res Board Can 29(6):975–983

Dash P, Walker ND, Mishra DR, Hu C, Pinckney JL, D'Sa EJ (2011) Estimation of cyanobacterial pigments in a freshwater lake using OCM satellite data. Remote Sens Environ 115 (12):3409–3423

Dawson K (2002) Fish kill events and habitat losses of the Richmond River, NSW Australia: an overview. In: Cooper JAG, Jackson DWT (eds) Proceedings of the ICS meeting, J Coastal Res Special Issue 36:216–221

DeMott WR (1989) The role of competition in zooplankton succession. In: Sommer U (ed) Plankton ecology. Springer, Berlin, pp 195–252

DeMott WR (1999) Foraging strategies and growth inhibition in five daphnids feeding on mixture of a toxic cyanobacterium and green alga. Freshw Biol 42(2):263–274
Dodds WK, Bouska WW, Eitzmann JL, Pilger TJ, Pitts KL, Riley AJ, Schloesser JT, Thornbrugh DJ (2009) Eutrophication of U.S. freshwaters: analysis of potential economic damages. Environ Sci Technol 43(1):12–19
Dokulil M, Teubner K (2000) Cyanobacterial dominance in lakes. Hydrobiology 438:1–12
Francis G (1878) Poisonous Australian lake. Nature 18:11–12
Frey DG (1955) Distribution ecology of the cisco in Indiana. Invest Indiana Lakes Streams 4:177
Fulton RS (1988) Resistance to blue-green algal toxins by Bosmina longirostris. J Plankton Res 10 (4):771–778
Geurts JJM, Smolders AJP, Verhoeven JTA, Roelofs JGM, Lamers LPM (2008) Sediment Fe:PO4 ratio as a diagnostic and prognostic tool for the restoration of macrophyte biodiversity in fen waters. Freshw Biol 53(10):2101–2116
Gomez JA, Alonso CA, Garcia AA (2011) Remote sensing as a tool for monitoring water quality parameters for Mediterranean Lakes of European Union water framework directive (WFD) and as a system of surveillance of cyanobacterial harmful algae blooms (SCyanoHABs). Environ Monit Assess 181(1–4):317–334
Gower J, King S, Borstad G, Brown L (2005) Detection of intense plankton blooms using the 709nm band of the MERIS imaging spectrometer. Int J Remote Sens 26(9):2005–2012
Havens KE (2007) Cyanobacteria blooms: effects on aquatic ecosystems. In: Hudnell HK (ed) Proceedings of the interagency, International Symposium on Cyanobacterial Harmful Algal Blooms (ISOC-HAB): state of the science and research needs. Springer, New York, pp 733–747
Hu C (2009) A novel ocean color index to detect floating algae in the global oceans. Remote Sens Environ 113(10):2118–2129
Hu C, He M-X (2008) Origin and offshore extent of floating algae in Olympic sailing area. Eos Trans AGU 89(33):302–303
Hu C, Chen Z, Clayton TD, Swarzenski P, Brock JC, Muller-Karger FE (2004) Assessment of estuarine water-quality indicators using MODIS medium-resolution bands: initial results from Tampa Bay. Florida Remote Sens Environ 93(3):423–441
Hupfer M, Hilt S (2008) Lake restoration. In: Jorgensen SE, Fath BD (eds) Encyclopedia of ecology. Elsevier, Oxford, pp 2080–2093
ILEC/Lake Biwa Research Institute (eds) (1988–1993) Survey of the state of the world's lakes, vols I–IV. International Lake Environment Committee, Otsu and United Nations Environment Programme, Nairobi
Jeppesen E, Jensen JP, Søndergaard M, Lauridsen T, Landkildehus F (2000) Trophic structure, species richness and biodiversity in Danish lakes: changes along a phosphorus gradient. Freshw Biol 45(2):201–218
Jones GD (ed) (1994) Cyanobacterial research in Australia (land and water). CSIRO, Australia, pp 731–915
Kahru M, Horstmann U, Rud O (1994) Satellite detection of increased cyanobacteria blooms in the Baltic Sea: natural fluctuation or ecosystem change? Ambio 23(8):469–472
Kahru M, Savchuk OP, Elmgren R (2007) Satellite measurements of cyanobacterial bloom frequency in the Baltic Sea: interannual and spatial variability. Mar Ecol Prog Ser 343:15–23
Klapper H (1991) Control of Eutrophication in Inland waters. Ellis Horwood, London, pp 1–337
Kutser T (2004) Quantitative detection of chlorophyll in cyanobacterial blooms by satellite remote sensing. Limnol Oceanogr 49(6):2179–2189
Lampert W (1981) Inhibitory and toxic effect of blue-green algae on Daphnia. Rev Ges Hydrobiol 66:285–298
Ludsin SA, Kershner MW, Blocksom KA, Knight RL, Stein RA (2001) Life after death in Lake Erie: nutrient controls drive fish species richness, rehabilitation. Ecol Appl 11(3):731–746
Margalef R (1968) Perspective in ecological theory. University Chicago Press, Chicago, pp 1–111

Mason CF, Bryant RJ (1975) Changes in the ecology of the Norfolk Broads. Freshw Biol 5 (3):257–270

Matthews MW, Bernard S, Robertson L (2012) An algorithm for detecting trophic status (chlorophyll-a), cyanobacterial-dominance, surface scums and floating vegetation in inland and coastal waters. Remote Sens Environ 124:637–652

Nürnberg GK (1995) The Anoxic factor, a quantitative measure of anoxia and fish species richness in Central Ontario lakes. Trans Am Fish Soc 124(5):677–686

Oyama Y, Fukushima T, Matsushita B, Kamiya K, Kobinata H (2015) Monitoring levels of cyanobacterial blooms using the visual cyanobacteria index (VCI) and floating algae index (FAI). Int J Appl Earth Obstet Geoinf 38:335–348

Pizzolon L, Tracanna B, Prósperi C, Guerrero JM (1999) Cyanobacterial bloom in Argentinean inland waters. Lakes Reserv 4:101–105

Pretty JN, Mason CF, Nedwell DB, Hine RE, Leaf S, Dils R (2003) Environmental costs of freshwater eutrophication in England and Wales. Environ Sci Technol 37(2):201–208

Robarts RS (1985) Hypertrophy, a consequence of development. Int J Environ Stud 25:167–175

Sayer CD, Davidson TA, Jones JI (2010) Seasonal dynamics of macrophytes and phytoplankton in shallow lakes: a eutrophication-driven pathway from plants to plankton? Freshw Biol 55 (3):500–513

Scheffer M, Jeppesen E (1998) In: Jeppesen E, Søndergaard M, Søndergaard M, Hristoffersen K (eds) Alternative stable states. Springer, New York, pp 397–406

Scheffer M, Hosper SH, Meijer ML, Moss B, Jeppesen E (1993) Alternative equilibria in shallow lakes. Trends Ecol Evol 8(8):275–279

Sculthorpe CD (1967) The biology of aquatic vascular plants. Edward Amold Publishers, London, pp 1–610

Seehausen O, van Alphen JJM, Witte F (1997) Cichlid fish diversity threatened by eutrophication that curbs sexual selection. Science 277(5333):1808–1811

Simis SGH, Peters SWM, Gons HJ (2005) Remote sensing of the cyanobacterial pigment phycocyanin in turbid inland water. Limnol Oceanogr 50(1):237–245

Smith VH, Schindler DW (2009) Eutrophication science: where do we go from here? Trends Ecol Evol 24(4):201–207

Søndergaard M, Phillips G, Hellsten S, Kolada A, Ecke F, Mäemets H, Mjelde M, Azzella MM, Oggioni A (2013) Maximum growing depth of submerged macrophytes in European lakes. Hydrobiologia 704:165–177. doi:10.1007/s10750-012-1389-1

Subramaniam A, Brown CW, Hood RR, Carpenter EJ, Capone DG (2002) Detecting Trichodesmium blooms in SeaWiFS imagery. Deep Sea Res Part II 49(1–3):107–121

Vincent RK, Qin X, McKay RML, Miner J, Czajkowski K, Savino J, Bridgeman T (2004) Phycocyanin detection from LANDSAT TM data for mapping cyanobacterial blooms in Lake Erie. Remote Sens Environ 89(3):381–392

Wilson MA, Carpenter SR (1999) Economic valuation of freshwater ecosystem services in the United States: 1971–1997. Ecol Appl 9(2):772–783

Wynne TT, Stumpf RP, Tomlinson MC, Warner RA, Tester PA, Dyble J, Fahnenstiel GL (2008) Relating spectral shape to cyanobacterial blooms in the Laurentian Great Lakes. Int J Remote Sens 29(12):3665–3672

Yu SZ (1995) Primary prevention of hepatocellular carcinoma. J Gastroenterol Hepatol 10:674–682

Zohary T, Breen CM (1989) Environmental factors favouring the formation of *Microcystis aeruginosa* hyperscums in a hypertrophic lake. Hydrobiology 178:179–192

Part III
Ecosystem Service and Socioeconomic Aspects with Special Reference to Biodiversity

Chapter 7
Utilization of Freshwater Fish Biodiversity as Income Source of Poor Rural People (Case Study in Pampangan Subdistrict of South Sumatra Province, Indonesia)

Dina Muthmainnah, Zulkifli Dahlan, Robiyanto H. Susanto, Abdul Karim Gaffar, and Dwi Putro Priadi

Abstract South Sumatra Province is covered by 1.1 million hectares of swamp land, which were considered as marginal land due to wet and muddy conditions as well as the presence of wild animals. During the wet season, the swamp ecosystem becomes a productive fishing ground, providing fish resource for neighboring communities. So this productive area constitutes a source of income for fishermen in the area. The current system categorizes the swamp area in Pampangan subdistrict into three types according to the source of water. Overall, the swamp is inhabited by 46 species of fishes and prawn. Type-2 swamp is populated by 31 species, while a number of 27 and 16 species were found in the type-3 and type-1 swamp, respectively. Type-2 swamp comprises the highest diversity. Resource allocation for fisheries is managed by the local government into 21 water bodies with artificial border. Each water body is exploited by a group of two to ten fishermen using various fishing gears, the most effective of which is a barrier with box trap (*lulung*). Using this fishing gear, the production might reach around 624,170 kg of fish per year. This gear seems to be a nonselective gear which catches 23 fish species. The other type, i.e., the longline (*rawai*), is the most selective gear which catches only seven carnivorous fish species. Data collected from this study indicates that fish yield is variable among water bodies from 15,000 kg per year in Lebung Asem to 220,900 kg per year in Rasau Jungkal, depending on ecological conditions in the swamp. Thus, the management of fishing activity would be very important in keeping fish biodiversity and in order to increase the fish yield; therefore, the income of fishermen might be increased.

D. Muthmainnah (✉) • A.K. Gaffar
Research Institute for Inland Fisheries, Jln. Beringin no. 8, Mariana-Palembang 30763, South Sumatra, Indonesia
e-mail: dina.gofar@yahoo.co.id

Z. Dahlan • R.H. Susanto • D.P. Priadi
Environmental Doctoral Programme, Sriwijaya University, Jln. Padang Selasa no. 524, Palembang 30139, South Sumatra, Indonesia

S.-i. Nakano et al. (eds.), *Aquatic Biodiversity Conservation and Ecosystem Services*, Ecological Research Monographs, DOI 10.1007/978-981-10-0780-4_7

Keywords Swamp • Freshwater fish • Biodiversity • Income • Rural people • Pampangan • South Sumatra • Indonesia

Introduction

Indonesian waters represent some of the rich biodiversity in the world including marine, fresh, and brackish water fish. Djajadiredja et al. (1977) estimated that there are about 4000 fish species in the Indonesian waters, at least 950 freshwater or brackish water species are recorded in western Indonesia and Borneo (Kottelat et al. 1993). Utomo et al. (2007) and Husnah et al. (2008) reported that there are 233 species of freshwater fishes in South Sumatra waters, grouped into 38 families and 111 genera. Recently, Muthmainnah et al. (2012) focused their study on the lowland swamp water of Pampangan subdistrict of South Sumatra, and they found that there were 44 species of fish and two species of freshwater prawn.

Lowland swamp is a type of dynamic ecosystem, which is alternately changed from aquatic to terrestrial ecosystem and vice versa. Swamp ecosystem has ecological functions such as water storage, habitat of wild fauna and flora, and also economical function such as fishing grounds, animal grazing fields, and developments into agriculture lands (Mitsch and Gosselink 1986).

The total area of lowland swamp in Indonesia is about 33 million hectares which are grouped into tidal swamp and nontidal swamp. In South Sumatra Province, there was 1.1 million hectares of nontidal lowland swamp (Sumsel in Figure 2005), which was considered as marginal land because the land was usually watered, wet and muddy, and habited by dangerous wild animals. During the wet season, swamp ecosystem becomes a productive fishing ground, providing fish resource as food for local community and also as commodities for fishermen to earn real income household. Some areas of deep swamp also develop into aquaculture sites (Gaffar and Muthmainnah 2011). Swamp management should focus on human needs within several ecological functions and ecosystem services.

Based on hydro-ecological characteristics, Pampangan's swamp area can be classified into three types. The main fishery activities by local people are fishing, and in some point culture fishery has been practiced. Fishing is performed either by individuals or groups of fishermen. Fishing license can be obtained through an auction system arranged by the local government. The auction winner is given a 1-year license to exploit one specific water body with artificial borders. The main issue in swamp fisheries is focused on conservation, wise use, and equity among all users.

This paper would discuss an aspect of swamp management as the case study in Pampangan subdistrict of South Sumatra, Indonesia, where local people work in the swamp ecosystem as small-scale fishermen to support their family life. Fishery resources in the swamp water are considered as important sources for food and income for many local villagers.

Ecological Characterization of the Swamp Ecosystems

The methods adopted in this study are field collection of samples and laboratory experiments. The samples were collected to identify the ecological conditions such as hydrology, water quality, vegetation, plankton, and fish diversity at several sampling points. According to sources of water, those swamps were divided into three types: (1) the swamp inundated by floodwater from Komering river which covers four villages, i.e., Tapus, Ulak Depati, Manggeris, and Pulau Betung; (2) the swamp with peat soils inundated by rainwater called Lebak Deling which covers three villages, i.e., Jungkal, Serdang, and Deling; and (3) the swamp inundated by both floodwater from Komering river and rainwater from Lebak Deling which covers three villages, i.e., Bangsal, Kuro, and Pulau Layang (Fig. 7.1).

The water level in those swamps was strongly influenced by precipitation rate, as shown in Fig. 7.2. High precipitation during the months of January to April and October to December causes higher water level of swamps in the subsequent months. And, in December, the average precipitation of 308.7 mm gave water level of the swamp of 169.3 cm, and at that time water from river flow into the lateral plain covered large areas of flood plain swamp.

Overall, there are 46 species of fishes and prawn in the swamp ecosystems. As shown in Table 7.1, the most diverse species (31) were found in type-2 swamp, while the least diverse species (16) were found in type-1 swamp. Type-3 swamp has a medium diversity of 27 species.

Taxonomically, the 44 species of fish were grouped into 15 families belonging to five orders, and two species of prawn were members of one family. Fish families consist of different numbers of species: Cyprinidae (16 species = 34.78 %), Channidae (6 species = 13.04 %), Siluridae (5 species = 10.87 %), and Anabantoidei (5 species = 10.87 %).

Eight species were found in all types of swamps, i.e., *Mystus nemurus*, *Channa striata*, *Cyclocheilichthys apogon*, *Cyclocheilichthys armatus*, *Pristolepis fasciata*, *Puntius lineatus*, *Osteochilus hasselti*, and *Trichogaster pectoralis*. There were six species of snakehead fish of genus *Channa* found in type 2 of swamp, while only one and two species were found in swamp of types 1 and 3, respectively. This finding indicates that snakehead fish of genus *Channa* might tolerate acidic black water in swamp with peat soils.

Utilization of the Swamp Ecosystems by Local People

Swamp areas have been considered providing ecosystem services which are important to support socioeconomic needs of the people living in the surrounding areas. Besides as fishing grounds mainly during wet season, the areas are also utilized as land for swamp rice cultivation, animal pasture, vegetable plots, and site for cage

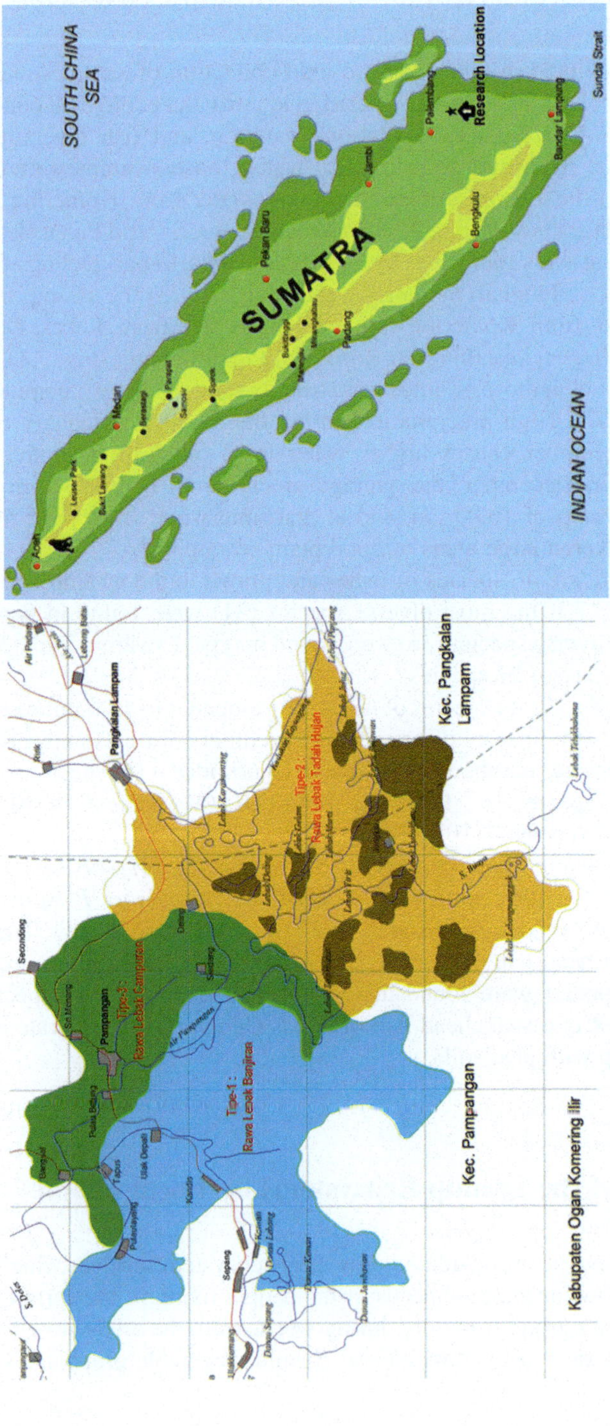

Fig. 7.1 Type of swamp accorded to sources of water in Pampangan subdistrict, Ogan Komering Ilir Regency, South Sumatra Province, Indonesia

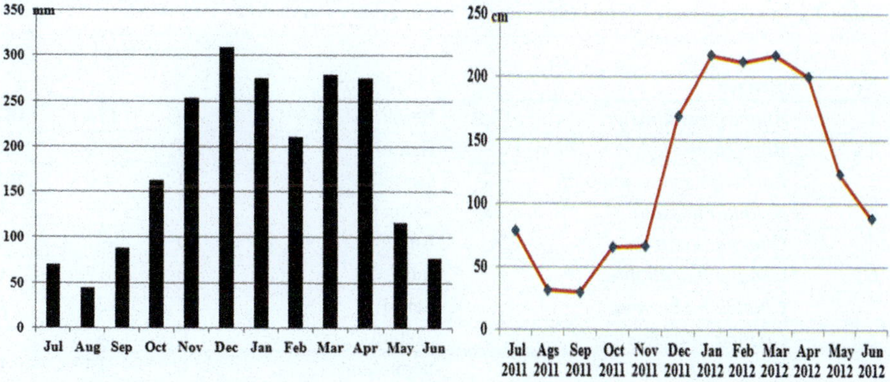

Fig. 7.2 Monthly average precipitation (2001–2011) (*left*) and water fluctuation (July 2011–June 2012) (*right*)

fish culture (Table 7.2). The survey was conducted to study the utilization of the swamp area in Pampangan subdistrict according to the type of swamp.

Water level in type-1 swamp was influenced by water level in the river. During the wet season, water level in the rivers rises and overflows into the swamp; thus, the water bodies become suitable for fishing ground. On the other hand, during the dry season, the water level lowers and flows back into the river causing major part of swamp to become dry land; thus, they can be utilized as rice fields or grazing fields for buffalos and cattle depending on the water level. Forty-four percent (44 %) of respondents reported that they used the land as rice fields during dry season.

Type-2 swamp, as found in three villages (Serdang, Deling, and Jungkal), was covered by quite deep water as a permanent water body, and the water has black color due to peat soils with acidic reaction (pH = 4.5). Water level fluctuated according to the variation of monthly precipitation. The swamp was covered with grasses and aquatic weeds; hence, the habitat of many species of fish which were collectively called black fish inhabited there. Twenty-six percent (26 %) of respondents reported that the areas were utilized as fishing ground by local fishermen using many kinds of fishing gears. During the dry season, some deeper water areas become swampy pools and allow the fishermen to catch more fish. A part of the swamp area becomes dry lowland where 32 % of respondents utilize the land for plots of short crops and 20 % of respondents utilize the land as buffalo grazing pasture. In this type of swamp, the dry period lasts only for about 3 months (during the peak of dry season).

Type-3 swamp keeps water from outflow of both river water and rain. The water reaction is acidic to neutral with grass vegetation. During the wet season, the swamp becomes a fishing ground, while during the dry season, 59 % of respondents utilize the area as rice field and 13 % of respondents utilize the area as buffalo grazing pasture.

Table 7.1 The number of fishes (individuals) occurred at swamp types

No.	Species	Local name	Swamp types		
			1	2	3
1	*Anabas testudineus*	Betok	0	173	157
2	*Barbichthys laevis*	Bentulu	0	0	26
3	*Barbodes schwanenfeldii*	Lampam	24	0	165
4	*Belontia hasselti*	Selincah	0	367	0
5	*Channa lucius*	Bujuk	0	431	0
6	*Channa marulioides*	Jalai	0	13	0
7	*Channa melasoma*	Serko	0	121	0
8	*Channa micropeltes*	Toman	0	85	39
9	*Channa pleuropthalmus*	Serandang	0	15	0
10	*Channa striata*	Gabus	120	540	197
11	*Clarias batrachus*	Lele	0	0	16
12	*Clarias nieuhofii*	Keli panjang	0	10	0
13	*Cyclocheilichthys apogon*	Keperas merah	44	311	526
14	*Cyclocheilichthys armatus*	Keperas putih	20	31	496
15	*Cyclocheilichthys enoplos*	Lumajang	12	0	0
16	*Hampala ampalong*	Tenggago	2	0	0
17	*Helostoma temminckii*	Sapil	0	456	384
18	*Kryptopterus schilbeides*	Lais kukur	0	44	0
19	*Kryptopterus apogon*	Lais muncung	16	0	91
20	*Kryptopterus cryptopterus*	Lais kaca	0	0	24
21	*Kryptopterus macrocephalus*	Lais tapah	12	29	0
22	*Labeo chrysophekadion*	Sihitam	1	0	4
23	*Macrobrachium rosenbergii*	Udang	0	6	2
24	*Metapenaeus brevicornis*	Udang serengkek	0	25	16
25	*Mystus nigriceps*	Berengit	123	0	0
26	*Mystus planiceps*	Baung	200	55	295
27	*Mystus wolffii*	Lundu	0	43	0
28	*Nandus nebulosus*	Setambun	0	465	0
29	*Notopterus notopterus*	Putak	0	0	25
30	*Osteochilus lineatus*	Tembelikat	0	0	16
31	*Osteochilus schlegelii*	Semuruk	0	42	0
32	*Osteochilus hasselti*	Palau	54	45	67
33	*Osteochilus microcephalus*	Kojam	0	0	55
34	*Oxyeleotris marmoratus*	Betutu	0	0	1
35	*Pangasius djambal*	Patin	0	0	12
36	*Parachela oxygaster*	Siamis	0	74	1
37	*Pristolepis fasciata*	Kepor/sepatung	64	379	73
38	*Pseudeutropius brachypopterus*	Riu	0	256	45
39	*Puntius hexazona*	Elang	24	56	0
40	*Puntius lineatus*	Kemuringan	124	661	150
41	*Puntius tetrazona*	Pirik cawang	0	24	0

(continued)

Table 7.1 (continued)

No.	Species	Local name	Swamp types		
			1	2	3
42	*Rasbora borneensis*	Seluang	0	290	0
43	*Tetraodon* sp.	Buntal	0	1	0
44	*Trichogaster trichopterus*	Sepat mata merah	0	0	303
45	*Trichogaster pectoralis*	Sepat siam	51	156	427
46	*Wallago leeri*	Tapah	0	15	0

Table 7.2 Percentage of people utilizing the lowland swamp as their household

No.	The activities	Percentage of people		
		Swamp type 1	Swamp type 2	Swamp type 3
1	Rice cultivation	44	4	59
2	Short crop	4	32	5
3	Capture fisheries	11	26	6
4	Fish culture	26	5	11
5	Raising buffalo swamp	5	20	13
6	Raising duck	7	4	6
7	Collecting wood	3	0	0
8	Collecting aquatic plant	0	5	0

Fisheries Activities

Data of fishing activities were collected by distributing questionnaires and direct interviews to 21 fishermen randomly selected within 21 water bodies. The result shows that each body is exploited by two to ten members of a fishermen group using many kinds of fishing gears. Usually the fishermen are members of local community surrounding villages. There was no special catching target. Some dominant species were *Channa lucius*, *Channa micropeltes*, *Channa pleuropthalmus*, *Channa melasoma*, *Nandus nebulosus*, *Anabas testudineus*, *Clarias batrachus*, *Helostoma temminckii*, *Puntius lineatus*, *Parachela oxygaster*, *Mystus nigriceps*, *Belontia hasselti*, *Pristolepis fasciata*, *Wallago leeri*, *Trichogaster pectoralis*, *Kryptopterus* sp., *Cyclocheilichthys apogon*, *Cyclocheilichthys armatus*, *Mystus planiceps*, *Barbodes schwanenfeldii*, *Macrobrachium rosenbergii*, and *Fluta alba*. Fishing activities were conducted from February to December, but each kind of fishing gear was operated in a specific period or season (Fig. 7.3).

In type-1 swamp, they used six kinds of fishing gears such as lift nets, filtering devices, pot traps, pole and lines, gill nets, and barriers with box trap, whereas in type-2 swamp, the gears they used were seines, lift nets, cast nets, longlines, pot

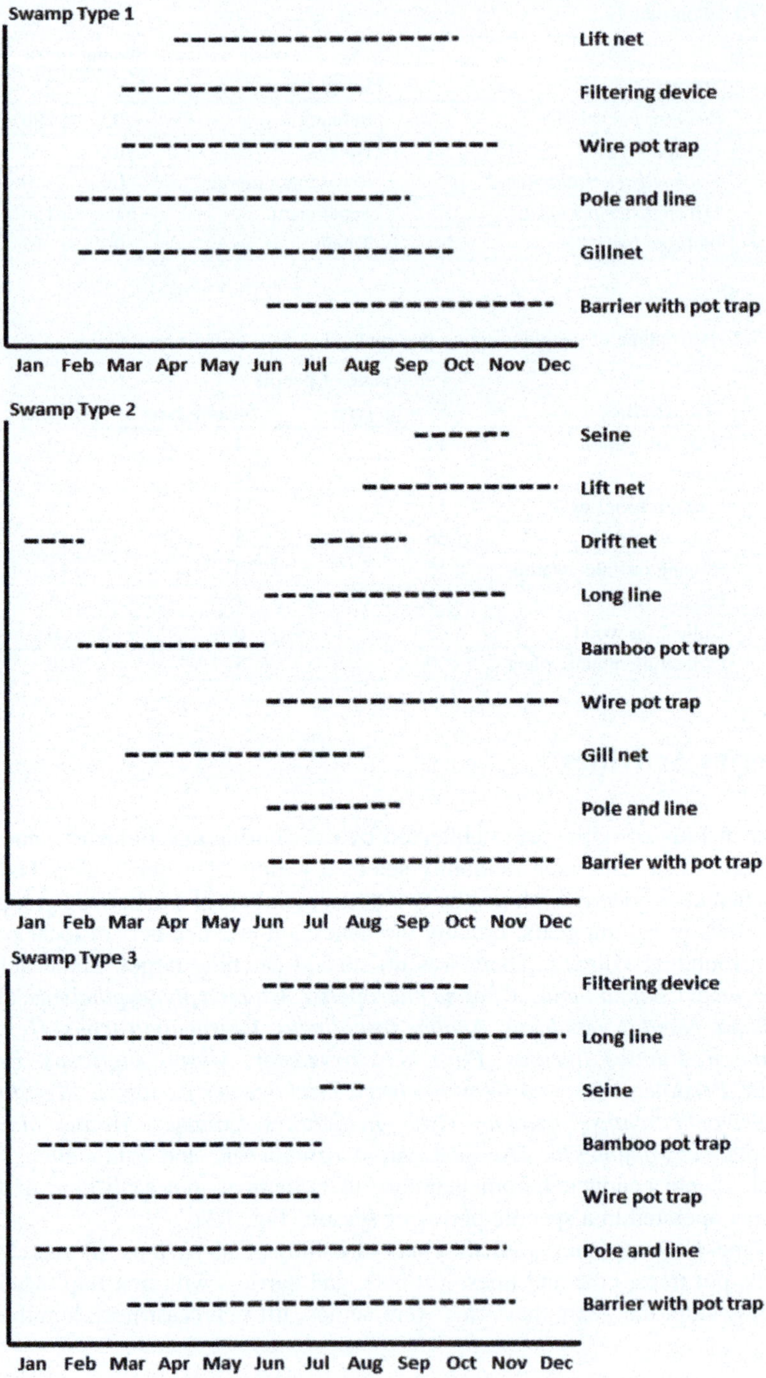

Fig. 7.3 Fisheries activity using different types of fishing gears operated according to dynamic of water level

Table 7.3 Data of fishing gear names and its total yield per year

Name of fishing gears	Total fishermen	Total yield (kg)
Barrier with pot trap (lulung)	102	624,170
Gill net (jaring)	73	114,480
Pole and line (tajur)	99	210,180
Wire pot trap (bangkirai)	93	249,680
Bamboo pot trap (bubu)	13	24,120
Seine (arat/kerakat)	10	5200
Longline (rawai)	73	96,960
Filtering device (tuguk/corong)	20	273,900
Others (cast net and lift net)	52	61,200
Total	535	1,659,890

traps, pole and lines, gill nets, and barriers with box trap. In type-3 swamp, the gears operated were filtering device, longline, seine, pot trap, wire pot trap, pole and line, and barrier with box trap.

This study shows that the most effective fishing gear is a barrier with pot trap (lulung) which catches around 624,170 kg/year. Barrier with pot trap (lulung) is a nonselective gear which catches 23 fish species, while a longline is the most selective gear which catches only seven carnivorous fish species (Table 7.3.). Another fishing gears which grouped as nonselective gears are seine and filtering device. Seine usually operated during the lowest water level where fish tend to concentrate in swamp pool causing almost all fish could be caught out. Filtering device usually operated during fast flow water where fish in moving against water current could be filtered by those gears. Regulation on fishing activity especially for nonselective gears seems to be necessary to support sustainable use of fish stock.

Fish yields were variable among water bodies from 15,000 kg/year in Lebung Asem to 220,900 kg per year in Rasau Jungkal (Table 7.4.). The fishes were marketed as live or fresh fish, and some species processed into salt fish, smoked fish, and fermented fish. Fish is marketed from fishermen to city market via collectors or middle men, but some fishermen directly sell their fish to consumers. According to the local prices, the fish could be grouped into three levels, the cheap fish less than IDR 15,000/kg of the average, the medium-price fish of IDR 15,000–30,000, and the high-price fish of more than IDR 30,000.

Fish culture is also practiced in swamp areas mainly using bamboo cages as fish enclosures. Local species of *Channa micropeltes* is the main cultured species followed by *Pangasius pleuropthalmus*. There were 1306 units of cages found in 13 villages around the swamp areas with yearly production of 200 kg per cage in average during the field observation.

Table 7.4 Data of swamp name and total yield per year

Name of swamp	Total yield (kg)
Lebak Deling	92,940
Lebung Asem	15,000
Kedukan Kiagung	17,880
Danau Jungkal	62,050
Murti Jungkal	44,350
Sebumbung Jungkal	54,300
Tiris Jungkal	76,800
Pinangboring Jungkal	24,810
Gabus Jungkal	172,980
Semunting Deling	87,780
Kuro	69,100
Muara Deles	6720
Lubuk Sekayan	25,800
Gelam Jungkal	36,900
Keliling Pulau Jungkal	145,700
Sengah Buye Jungkal	62,300
Rasau Jungkal	220,900
Camang Ulak Depati	136,080
Lepok Ulak Depati	97,650
Sematang Bunder Ulak Depati	74,400
Perompong Ulak Depati	182,250
	1,706,690

Conclusions

The swamp area in Pampangan subdistrict is inhabited by 46 species of fishes and prawn. The swamp is grouped into three types depending on the source of water. Type-2 swamp, with a quite deep permanent water level, comprises the most diversity in fish and prawn species. The swamp areas have been considered as a part of ecosystem which is important in supporting the socioeconomic need of the people in the surrounding areas. The management of resources allocation for fisheries is managed by the local government in 21 water bodies with artificial borders. Data collected for this study indicates that the fish yields are variable among water bodies from 15,000 kg per year in Lebung Asem to 220,900 kg per year in Rasau Jungkal depending on water levels in the swamp and types of fishing gears. The most effective fishing gear was a barrier with pot trap (lulung). The study concluded that the management of fishing activity would be very important in keeping fish biodiversity and in order to increase the fish yield; therefore, the income of fisherman might be increased.

Acknowledgments The present study was funded by the Ministry of Marine and Fisheries, Republic of Indonesia, throughout Research Institute for Inland Fisheries in 2012.

References

Djajadireja R, Fatimah S, Arifin Z (1977) Economical important fish species. Ditjen Perikanan. Departemen Pertanian, Jakarta

Gaffar AK, Muthmainnah D (2011) Inland fisheries management business. Proceeding 8th Indonesian conference on Indonesian inland waters. Research Institute for Inland Fisheries. Research Center for Fisheries Management and Fish Resources Conservation. The Agency for Marine and Fisheries Research and Development. Palembang. ISBN. 978-602-8380-05-8. p 63–68

Husnah, Nurhayati E, Suryati NK (2008) Diversity morphological characters and habitat of fish in 204 Musi River drainage area. Research Institute for Inland Fisheries, South Sumatra, 440 p

Kottelat M, Whitten AJ, Kartikasari SN, Wirjoatmojo S (1993) Freshwater fishes of western Indonesia and Sulawesi. Periplus Edition-EMDI Project, Jakarta

Mitsch WJ, Gosselink JG (1986) Wetlands. van Nostrand Reinhold Company, New York, 539 pp

Muthmainnah D, Dahlan Z, Susanto RH, Gaffar AK, Priadi DP (2012) The distribution of fishes in Lebak Pampangan swamp south Sumatra Province. Presented in 1st international seminar Indonesian Ichthyological Society, The Faculty of Marine Science and Fisheries, Hasanuddin University and Indonesian Ichthyological Society, Makasar, 12–13 June 2012, 13 pp

Sumsel in Figure (2005) Collaboration between Bappeda South Sumatra with BPS South Sumatra Province, Palembang. Sumatera Selatan Dalam angka, Badan Pusat Statistik, Provinsi Sumatera Selatan, Palemban

Utomo AD, Muflikhah N, Nurdawati S, Rahardjo MF, Makmur S (2007) Ichthyofauna in Musi River South Sumatra. Research Institute for Inland Fisheries, Jakarta, 380p

Chapter 8
Why People Visit Zoos: An Empirical Approach Using the Travel Cost Method for the Higashiyama Zoo, Nagoya, Japan

Ryo Kohsaka, Kaho Naganawa, and Yasushi Shoji

Abstract This paper evaluates the cost and other benefits accumulated by the Higashiyama Zoo of the Nagoya Higashiyama Zoo and Botanical Gardens due to recent renovations. We examine changes in values and appreciation of visitors. The simple method of travel cost is applied to study consumer surplus of the zoo, which preliminarily indicates a surplus of 4566 yen.

Keywords Exhibitions • Consumer surplus • Zoo • Budgetary constraint • Travel cost

Introduction

Worldwide, there are some 10,000 zoos with more than 700 million visitors (WAZA 2005). In Japan, as of April 2013, there are as many as 86 registered zoos under the Japan Association of Zoos and Aquariums or JAZA (these figures do not include the private zoos).

The futures of Japanese zoos are not as rosy as their quantity indicates. Fiscal perspectives for most of the zoos in Japan are grim due to fewer visitations, possibly associated with decreasing numbers in younger generations and a rapidly aging population. The majority of Japanese zoos are owned or financially supported by municipalities, which face budgetary constraints as tax income decreasing due to changes in populations.

R. Kohsaka (✉)
Graduate School of Human and Socio- Environmental Studies, Kanazawa University, 1 Kakuma, Kanazawa 920-1192, Ishikawa, Japan
e-mail: kohsaka@hotmail.com

K. Naganawa
School of Rural Development, Kanazawa University, 1 Kakuma, Kanazawa 920-1192, Ishikawa, Japan

Y. Shoji
Research Faculty of Agriculture, Hokkaido University, Kita 9 Nishi 9, Kita-ku, Sapporo 060-8589, Hokkaido, Japan

S.-i. Nakano et al. (eds.), *Aquatic Biodiversity Conservation and Ecosystem Services*, Ecological Research Monographs, DOI 10.1007/978-981-10-0780-4_8

There are initial signs of such constraints. Maruyama Zoo at Sapporo recently implemented a questionnaire asking their citizens the merits of reintroducing elephants from abroad and the readiness for bearing the costs (Sapporo City 2012). The result implied that the citizens appreciated the option of the reintroduction, replying that the children will benefit from the exhibition of elephants, but, simultaneously, the citizens were also concerned about the costs of such actions, ranking this response as the second most frequent from the range of answers. The need for such questionnaires, between budgetary balances and services the facilities can provide, is likely to continue due to changing municipal incomes and priorities. Questionnaires as these demonstrate a field of applied research that has yet to be explored.

Despite such practical needs, surprisingly limited literature is found that analyzes zoos using economic evaluations and empirical datasets. The main focus of existing literatures is either on roles of zoos or quantitative analysis on visitors, such as preferred animals, methods of display, or educational effectiveness. To the knowledge of the authors, empirical analysis of zoos with economic approaches is rather limited. The purpose of this paper is to economically evaluate a zoo with empirical data in order to support decision makings in the changing context and priorities of management and budgetary constraints.

Literature Review

The existing literature can be categorized into zoos and visitors. The research methods of Forst (2011:235) mentions the "supply side," examining what the zoos are doing and why. His approach usefully demonstrates that zoos are not passive but are responding to criticisms and rethinking their roles in a strategic way. The other approach is the "demand side," which examines choices, expectations, and motivations of the tourists (ibid.).

The first group of literature on zoo research focuses mainly on education and conservation. Historically, zoo research discusses how zoos should be organized or how animals should be treated and relate mainly to the fields of ethics, sociology, or veterinary sciences. The zoos themselves are traditionally the topic of debates, ranging from morals, educational programs, animal rights, and human welfare. These debates reflect the social movements and criticism against the zoos in both Europe and Northern American since the 1980s.

The functions of the zoos date back to Curtis (1968) in the classical textbook of *Zoological Park Fundamentals* which emphasizes the multifunction of the zoo in conservation, education, research, and entertainment. More recently, a number of scientists and practitioners discussed the pros and cons of zoos as institutions and formed three categories: education, science (including conservation), and entertainment (Mason 2011; Shackley 1996; Bostock 1993; Jamieson 1995). Some researchers argue for different forms of animal exhibition (a "non-zoo" model) where "visitors would witness volunteers or volunteers themselves in the rehabilitation and care of these animals" or live streaming video of wild animals on the internet (Milstein 2009: 28). Some authors questioned the role of zoos, particularly in relation to entertainment, from the viewpoint of animal rights and welfare.

More recently, the focus is less on the existence of the zoos but rather on how zoos could be of better service to the society. Based on these reviews, the question raised is how can there be a balance among the different functions of the zoos, i.e., education, entertainment, research, and conservation (van Linge 1992; Linke and Winter 2011). Kawata (2011) questions the large ratio of large mammals in exhibitions that are presumably based on an anthropogenic point of view. Thus, Hanson (2002) points out that the zoo is a combination of "science and showmanship" (Hanson 2002). There is a rich amount of literature in this first group of zoo research that catalyzes the field of "human-animal relationships." Yet these debates need to be considered under the limitations of budgetary constraints and resources and actions must be prioritized. This paper aims to partially fulfill the information gap by evaluating a zoo under construction reform.

In the second group of literature, or so-called researches in the "demand-side" field, zoo visitations are evaluated mostly with quantitative approaches. In this research, the underlying motivations of the zoo visitors are analyzed. The roles of zoos are categorized as educational, scientific, or recreational. Most quantitative approach to zoo visitations explores the preferences of the animals on display or compares the preferences among visitors and non-visitors with different social attributes.

There are studies using the term "experience economy," but their actual focus is on methods to design, stage, or dramatize touristic experiences (cf. Pine and Gilmore 1999). There is research comparing the awareness of visitors to botanical gardens and zoos (Ballantyne et al. 2008). Raede and Waren (1996), for example, compared visitors and non-visitors on their perception of zoos and that of animal welfare for the Edinburgh Zoo. The results indicated that the public outside of the zoo tended to view zoo animals negatively as bored and sad with variation by age and sex, while the actual visitors tended to be more positive about the role of zoos and animals.

This paper analyzes the preferences of zoo visitors through quantitative data, such as their travel costs. Thus, it aims to complement the existing literature with data on travel costs that were paid by the visitors.

Case

The Higashiyama Zoo is the second most visited zoo in Japan annual, next to Ueno Zoo in Tokyo, with 2.18 million visitors in 2010. It is located within the district of Nagoya City, with direct access from the urban subway system. As the formal name Nagoya Higashiyama Zoo and Botanical Gardens indicates, it is a combination of a zoo and a botanical garden. The area covered is 59 ha (the zoo part has 32 ha). The zoo was established in 1937 and contains more than 125 species, including the popular koalas and the famous collection of *medaka* or the Japanese rice fish (*Oryzias latipes*). Since 2009, it is undergoing a revised reform plan to renovate its facilities.

Higashiyama Zoo is currently undergoing a major renovation and thus provides us with a unique opportunity to explore the balance between benefits and costs with an economic evaluation. Is it worthwhile for the zoo to undergo renovations? What are the consumer surpluses (arising from such renovations)? Are people who know about the renovations likely to visit the zoos with different frequency? What are the main purposes or motivations of the visitors to the Higashiyama Zoo? Is it considered as an attraction for entertainment or for educational purpose to expose children to zoos?

Do the changes in the design of the Higashiyama Zoo affect the behavior of the visitors?

Method

Structured questionnaire survey was conducted on site in Higashiyama Zoo for the period of August 13–17, 2012. The period overlaps with summer holiday season in Japanese school calendar which presumably has an implication for the results. Visitors to the Higashiyama Zoo were asked face-to-face to fill in the questionnaire.

The travel cost method (TCM) has traditionally and widely been used to estimate recreational values (Ward and Beal 2000; Haab and McConnell 2002; Champ et al. 2003). Single-site models are useful when the goal is to estimate the total use or access value of a site (Parsons 2003). The simplest form of the single-site model is as follows:

$$r = f(tc, s)$$

where r is the number of trips to a zoo, tc is the travel cost for the zoo, and s is the socioeconomic variables.

People in the zoo's neighborhood face lower travel costs, whereas people who live far away from the zoo incur higher travel costs. Thus, the parameter of tc is assumed to be negative, under a ceteris paribus condition. In general expression, consumer surplus (CS), which is visitor's willingness to pay for net travel cost, is defined in the area under the recreation demand function between visitors' travel cost tc^* and choke price tc_{choke}, at which trips fall to zero:

$$CS = \int_{tc^*}^{tc_{choke}} f(tc)\, dtc$$

In this study, the recreation demand function is specified with the number of trips per *** as the dependent variable. The dummy variables of existence or nonexistence of alternative recreation sites *alt* and the knowledge about repairmen of the zoo *rpr* as well as *tc* are independent variables.

Due to the nature of the onsite sampling survey, the estimation of the recreation demand function includes three issues: nonnegative integer-dependent variables, truncation, and endogenous stratification (Shaw 1988). To cope with the biases, truncated Poisson or negative binomial regressions are often used (Creel and Loomis 1990, 1991; Hellerstein and Mendelsohn 1993).

The probability of observing an individual visiting the zoo during *** is

$$\Pr(r) = \frac{e^{-\lambda} \cdot \lambda^r}{r!}, \quad r = 1, 2, \ldots$$

where λ is both the mean and variance of the distribution, and usually specified as $\lambda = \alpha + \beta_{tc} \cdot x_{tc} + \beta_{alt} \cdot alt + \beta_{rpr} \cdot rpr$, which is vector of dependent variables and their parameters. The *CS* of individual *n* is

$$CS_n = -\frac{\lambda_n}{\beta_{tc}},$$

where λ_n is the individual *n*'s expected value of trips to the zoo. Considering the truncation at one trip and endogenous stratification (people who go to the zoo frequently are more likely to be sampled than people who go to the site occasionally) above takes the form

$$\Pr(r|r > 0) = \frac{e^{-\lambda} \cdot \lambda^{r-1}}{(r-1)!},$$

The negative binomial version of revised model is also derived in the same fashion.

Results

A total of 78 individuals responded to the request and data was collected. Based on the model, the consumer surplus was estimated to be 4566 yen, and the zoo's renovation had a significant impact on both the consumer surplus and frequency of visits.

The preliminary results indicated that visitor's knowledge of the ongoing renovations of the Higashiyama Zoo affected the frequency of visitations. In other words, when visitors knew about the renovations in the zoo, they were likely to visit the zoo more frequently resulting in a higher consumer surplus on average of 7404 yen. It is estimated that the consumer surplus added 2838 yen through the renovation of the zoo (a difference of the two amounts 7404 yen and 4566 yen) by comparing those who did not list the aquarium as alterative.

Table 8.1 Poisson and negative binomial model considering endogenous stratification

	Poisson model		Negative binomial model	
	Normal	Truncation	Normal	Truncation
Constant	1.3464***	1.0965***	1.5816***	1.089***
Travel cost ($*10^{-3}$)	−0.9202***	−1.0135***	−0.5945***	−0.7212***
Alternative destination	−1.5070***	−1.7750***	−0.7500**	−1.7868***
Prior information on renovation	1.811***	2.0356***	1.0986***	2.0464***
Alpha ($=1/\theta$)			0.9570***	1.3714***
Number of observation	78	78	78	78
Maximum likelihood	−249.7375	−490.8719	−200.6790	−175.2718
Consumer surplus for one traveler (JPY)	3579	3249	3249	4566

$^{***}p < 0.01$, $^{**}p < 0.05$

The model estimates four results (Table 8.1). The column negative binomial model with truncation is adopted, and the results have the least figure of "maximum likelihood." As the travel cost increases, the frequency of visit will naturally decrease. The figures of alternative destination will decrease in a similar manner. On the other hand, those who had prior information about the renovation had a higher frequency of visits.

In the estimate model, visitors with "prior information on renovation" will visit 3.293 times on average that is 2.0464 higher than the overall average. Based on this result, the consumer surplus is estimated to be 7404 yen (with negative binomial model with truncation).

In conclusion, the renovation information will increase the frequency of the visit and increase the consumer surplus. The difference (2838 yen) between 7404 and 4566 yen is estimated to be the impact of renovation. Further research is required to see additional changes and benefits due to the renovations so that fiscal outlooks of zoos do not remain grim in the future. Preliminarily, it seems renovations can increase zoo's costs and benefits when visitors are aware of the changes. This can aid budgetary decision making of municipally funded zoos, providing similar factors (i.e., access from the urban subway system), in that widespread information dissemination of changes may prove to be cost effective.

References

Ballantyne R, Packer J, Hughes K (2008) Environmental awareness, interests and motives of botanic gardens visitors: implications for interpretive practice. Tour Manage 29:439–444

Bostock SSC (1993) Zoos and animal rights; the ethics of keeping animals. Routledge, London

Champ PA, Boyle KJ, Brown TC (eds) (2003) A primer on nonmarket valuation. The economics of non-market goods and resources. Kluwer, Dordrecht

Creel MD, Loomis JB (1990) Theoretical and empirical advantages of truncated count estimators for analysis of deer hunting in California. Am J Agric Econ 72(2):434–441

Creel MD, Loomis JB (1991) Confidence intervals for welfare measures with application to a problem of truncated counts. Rev Econ Stat 73(2):367–274

Curtis L (1968) Zoological park fundamentals. American Association of Zoological Parks and Aquariums, Washington, D.C., 80 pp

Frost (2011) Zoos and tourism in a changing world. In: Frost W (ed) Zoos and tourism: conservation, education, entertainment? Channel View Publications, Bristol, pp 227–235

Haab TC, McConnell KE (2002) Valuing environmental and natural resources. Edward Elgar, Northampton

Hanson (2002) Animal attractions; nature on display in American zoos. Princeton University Press, Princeton

Hellerstein D, Mendelsohn R (1993) A theoretical foundation for count data models. Am J Agric Econ 75(3):604–611

Jamieson D (1995) Against the zoos. In: Singer P (ed) Defense of animals. Blackwell, Hoboken, pp 108–117

Kawata K (2011) Romancing the celluloid nature: a review of American zoo exhibit, part. Der Zool Garten 80:239–253, http://dx.doi.org/10.1016/j.zoolgart.2011.08.002

Linke S, Winter C (2011) Conservation, education or entertainment; what really matters to zoo visitors? In: Frost, W (ed) Zoos and tourism: conservation, education, entertainment? Channel View Publications, Washington, D.C., pp 69–84

Mason P (2011) Zoos and the media. In: Frost W (ed) Zoos and tourism: conservation, education, entertainment? Channel View Publications, Washington, D.C., pp 189–203

Milstein T (2009) "Somethin' Tells Me It's All Happening at the Zoo": discourse, power conservationism. Environ Commun: J Nat Cult 3(1):25–48, http://dx.doi.org/10.1080/17524030802674174

Parsons GR (2003) In: Champ PA, Boyle KJ, Brown TC (eds) The travel cost model. A primer on nonmarket valuation. Kluwer, Dordrecht, pp 269–329

Pine JP, Glimore JH (1999) The experience economy; work is theater & every business a stage. Harvard Business School Press, Brighton

Raede LS, Waren NK (1996) The modern zoo: how do people perceive zoo animals? Appl Anim Behav Sci 47:109–118

Sapporo City (2012) Examination of introducing elephants at Sapporo Maruyama zoo. The first questionnaire to the citizens (in Japanese). http://www.city.sapporo.jp/zoo/topics/zouusiikutyousa.html

Shackley M (1996) Wildlife tourism. International Thomson Business Press, London

Shaw D (1988) On-site samples' regression: problems of non-negative integers, truncation, and endogenous stratification. J Econom 37(2):211–223

Van Linge JH (1992) How to out- zoo to zoo. Tour Manage 13:115–117

Ward FA, Beal DJ (2000) Valuing nature with travel cost models: a manual. Edward Elgar, Northampton

WAZA (2005) Building a future for wildlife- the world zoo and aquarium conservation strategy. WAZA, Gland

Chapter 9
Tourist Perceptions of Traditional Japanese Vegetable Brands: A Quantitative Approach to Kaga Vegetable Brands and an Information Channel for Tourists at the Noto GIAHS Site

Ryo Kohsaka, Mitsuyuki Tomiyoshi, and Hikaru Matuoka

Abstract Brands for local vegetables in various regions in Japan are created and entering competitions in the market. In this paper, we analyze one type of traditional vegetable, *Kaga*, which is named for one of the two subregions that constitute Ishikawa Prefecture, situated on the Sea of Japan. The Kaga brand has existed since the postwar era. Our analysis focuses on consumer perception and the limiting factors of the Kaga brand. For our research, we conducted face-to-face interviews with tourists at local food markets, using a structured questionnaire. We interviewed 378 Japanese respondents at the Noto Shokusai Market (*Noto Shokusai Ichiba*) and Wajima Morning Market (*Wajima Asaichi*), both popular tourists' destinations. In the questionnaire, we analyzed 12 parameters, including place of residence, age, gender, type of shop, information sources (television, Internet, friends' word-of-mouth advertising, guidebooks, posters, etc.), and purchase amounts. Specifically, we analyzed which channels and media as information sources on traditional vegetables influenced consumer choices. Analytical results indicate that the Internet, friends, guidebooks, posters, purchase amount, and age were not significant factors influencing purchase choice, while more conventional communications channels such as shops, television, and newspaper were more influential.

Keywords Traditional Japanese vegetables • Branding • Advertising effectiveness • Kaga vegetables • Tourism

---

R. Kohsaka (✉) • M. Tomiyoshi
Graduate School of Human and Socio-Environmental Studies, Kanazawa University, Kakuma, Kanazawa 920-1192, Ishikawa, Japan
e-mail: kohsaka@hotmail.com

H. Matuoka
Institute of Science and Engineering, Kanazawa University, Kakuma, Kanazawa 920-1192, Ishikawa, Japan

S.-i. Nakano et al. (eds.), *Aquatic Biodiversity Conservation and Ecosystem Services*, Ecological Research Monographs, DOI 10.1007/978-981-10-0780-4_9

Introduction

An increasing number of municipal authorities in Japan are creating market brands for their locally produced vegetables, as part of efforts toward regional revitalization and conservation of local traditions and history. Vegetables are produced by farmers and sold to consumers through retailers, similar to the distribution and marketing processes for other consumer products. The direct sale by the producers is increasingly common, often established alongside the road as local direct sale shops (*Chokuhanjo* or *Michi no Eki* in Japanese). Sales of one-to-one OTC (over the counter), at the centralized auction market for vegetables, are increasingly used, rather than the auction bid for the vegetables. In order to succeed in such sales, branding production site through brand strategy of the sites or geographical indications of products become critical. The Noto area is designated as Globally Important Agricultural Heritage Systems (GIAHS) in 2011 by the Food and Agriculture Organization of the United Nations (FAO). The reason of designation was the rich biological diversity in the peninsula and their interlinkage with agricultural systems symbolized by the terms satoyama and satoumi.

In such processes, advertising transmits information on production, lifestyle images, and other consumer concepts the marketer wishes to associate with the products and services including tourism or traditional landscapes. The impacts of advertising differ from new to existing products and media context (Janssens and De Pelsmacker 2005; Buttle 1998).

For Japanese vegetables, established brands include Kyo (produced in Kyoto) and Kaga (produced in Kaga region of Ishikawa Prefecture). Newer brands include Edo (Tokyo) and Nanihwa (Osaka), among others. The established brands enjoy national reputations, are served at restaurants nationwide, and are used in a variety of local cuisines. These brands promote a narrative about the history of their seeds and vegetables since the Edo period, from 1603 to 1868 (Matsushiata 2007). Literature has been published about the branding and promotion of individual Kaga vegetables such as the sweet potato (Yamamoto 2004; Nakamura and Niizawa 2006; Ishikawa Yasaikaki Research Group 2007; Nishikawa 2010). The prefecture and city of Kyoto are actively promoting the Kyo vegetable brand for nationwide consumption and retailing, and its rather “high-class” brand image is associated with nobles and history (Aoya 2010). Such differences in consumer groups are analyzed for organic vegetables in the American context, which indicated that age, higher income, and education levels are influencing factors for consumption of organic vegetables (Dettmann and Dimitri 2009).

The Agricultural Production Association of Kanazawa City listed 15 items under the Kaga vegetable brand as of June 2014 (Japanese name and/or English name, if available, and scientific name): gorozima kintoki (sweet potato, *Ipomoea batatas*), kinjiso (*Gynura bicolor*), Kaga tsurumame (hyacinth bean, *Lablab purpureus*), heta-murasaki-nasu (eggplant, *Solanum melongena*), bamboo shoots (*Phyllostachys heterocycla*), Kaga renkon (lotus root, *Nelumbo nucifera*), gensuke daikon (radish, *Raphanus sativus*), Utsugi Akagawa amaguri kabocha (squash, *Cucurbita maxima*), futatsuka karashina (leaf mustard, *Brassica juncea*), Kaga huto kyuri (Kaga cucumber, *Cucumis sativus*), Kaga ippon negi (Kaga welsh onion, *Allium fistulosum*), aka zuiki (taro stem, *Colocasia esculenta*), kuwai (three-leaf arrowhead, *Sagittaria trifolia*), and Kanazawa shungiku (crown daisy, *Chrysanthemum coronarium*).

For kinjiso, methods for annual production are well established and various new products are being developed. Certain products only have seasonal demand, such as kuwai for New Year celebrations (Ishikawa Vegetable and Flower Research 2007). By certifying traditional and local vegetables in the community, the products serve as advertising signals for consumers, in addition to continuity of technology, and maintenance of varieties (Yamamoto 2004).

There are challenges for the brand as well. The Kaga vegetables are certified exclusively for those items produced in the Kanazawa City area. Sweet potato (*gorojima kintoki*) production is not keeping up with increasing demand, resulting in opportunity costs for producers (Saito 2010). Meanwhile, efforts are underway to promote consumption. Restaurants and retailers have the option of registering as certified "Kaga Vegetable Shops" in the Kanazawa City area. Registered restaurants include those that serve Western and Chinese cuisine, in addition to traditional Japanese cuisine.

Among the 15 registered Kaga vegetables, production volume varies significantly; some items are grown in relatively large quantities, while others are barely enough to continue production. For the latter, the volume is insufficient to make production competitive. The volume is limited because of the limited production scale in farmers and geographical locations. Furthermore, the traditional vegetables are frequently more unstable in production volumes, vulnerable to climate and diseases, compared to the commercial hybrid varieties. These "minor crops" have the risk of disappearing. For such vegetables, one strategy is to establish long-term relationships with consumers, by branding the product with histories and stories relating to the specific vegetable (Otani and Saito 2008).

There are two reasons why this study has chosen to focus on consumer awareness of Kaga vegetables, in the Noto area. First, there is an existing dataset for Kaga vegetables in terms of their recognition and awareness among tourists and restaurants in Kanazawa (DBJ and Kohsaka Lab 2014). Therefore, our research in Noto will enable comparison with Kanazawa. Second, the Noto Shokusai Market (in Nanao City) and Wajima Morning Market (in Wajima City) where interviews will take place are important food-related tourist spots.

Review of Existing Literature

In this section, we review online advertising and online and personal communications as information channels relating to local and traditional vegetables.

Challenges and Profitability of Branding Traditional Vegetables

Since the early 1990s, trends have moved toward increasing consumption in restaurants and ready-made meals, while the use of fresh foods at home has declined. Furthermore, consumers have become more price conscious due to the economic recession; other trends include branding of agricultural products and industry efforts to sell more value-added products (Hazumi 2002: 2). There are other attempts to revitalize the local community by branding a single farm product at the local level (Savitri 2008). These movements to brand the agricultural products utilize intellectual property including trademarks and patents, enabling the local community to use the brand exclusively. By allowing purchase exclusively in the produced regions, the place of the production provides added value as tourist destinations (Saito 2010, 2011). As indicated in Table 9.1, in Kyoto, Kaga, or other areas, various stakeholders, such as farmers, retailers, municipalities, and agricultural associations, gather to vitalize the region as a whole in Japan.

In the example of Kyoto, the Kyoyasai vegetable brand was established by a cooperation of stakeholders, including chefs, farmers, academics, and the government (Rath 2014). In the United Kingdom and Italy as well, there have been reports that governments see the potential for vegetable brands for regional revitalization (O'Neill 2014). As for the Kaga brand, the name value of the brand is less established than Kyoyasai, but it is the second best well-known branded vegetable with rather limited amount of production (DBJ and Kohsaka Lab 2014). The development path, awareness, and communication channels of the Kaga brand by the consumers are critical for other traditional vegetable brands and local communities.

According to American Marketing Association quoted by Kotler (2000), brand is a "name, term, sign, symbol or design, or a combination of them intended to identify the goods and services of one seller or group of sellers and to differentiate them from those of other sellers." Marketing associations in America also follow this definition and use local brands to differentiate products.

The Regional Development Bank, DBJ, and researchers interviewed tourists and shop owners about what food items they found most appealing within the Ishikawa Prefecture and found that branded vegetables, including the Kaga brand, were second only to the popular local fish species (DBJ and Kohsaka Lab 2014). In the spring of 2015, an extension of the bullet train railway systems is expected to improve access to Kanazawa directly from Tokyo, a distance of about 650 km by rail. This transformation of the transportation system is expected to increase the number of tourists arriving here implying more consumptions and sales of traditional vegetables at the local sites.

From the results of DBJ and Kohsaka Lab (2014), it is clear that the Kaga vegetable brand is well known as a whole, but individual items are less known. Their study also found that service and communication provided at individual shops are important. Their findings suggest that the service sector, including restaurants,

Table 9.1 Characteristics of traditional vegetables in various regions of Japan

Branded vegetables	Definition	Number of items	Certification organization	Comment
Edo traditional vegetable (Edo vegetable)	Traditional vegetables cultivated from Edo period and vegetables that emerged between Meiji and mid-Showa period (1955 to 1964) through horticultural improvements	40	Edo Tokyo Vegetable Promotion Council in JA Tokyo Chuo	
Kyoto traditional vegetable	Introduced before Meiji period and produced in Kyoto Prefecture	40 (including extinct items)	Agriculture, Forestry and Fisheries Department, Kyoto Prefectural Government	27 items including Kyoyasai certified as Kyo brand products
Kaga vegetable	Cultivated before 1945 and still in production today, mainly in Kanazawa	15	Kanazawa City Branding Association of Agricultural Products	
Yamato traditional vegetable	Items documented to have been produced before World War II	18	Agriculture, Forestry and Fisheries Department, Nara Prefectural Government	Yamato vegetables consist of "Yamato traditional vegetables" and "Yamato kodawari vegetables"
	Items with distinctive taste, flavor, and morphology, derived from unique cultivation style based on local history and culture			
Noto traditional vegetable	Vegetables cultivated and used in traditional dietary culture in Noto region	6	Noto Vegetable Promotion Council (office hosted by Ishikawa Prefectural Government)	Noto vegetables consist of "Noto traditional vegetables" and "Noto specialty vegetables (7 items)"
	Generally cultivated more than 30 years			
	Developed by a farmers' group			
	Distinctive features			

needs to provide information in ways that distinguish traditional from nontraditional vegetables, particularly for individual vegetables. In addition to restaurants, the promotion of traditional vegetables is diversifying, including direct advertising by farmers, such as point-of-purchase displays (e.g., posters placed near the product on display), and comments on the Internet.

Reception of Consumers and Online Communication

One study points out that the place of residence is a critical factor in purchasing behavior and attitude toward vegetables (Kobayashi 2008). Studies have covered a wide variety of issues from the marketing of universities (Kuribayashi 2008) to paper diapers (Goto 2011). In the case of paper diapers, oral communications with friends and acquaintances ranked higher in importance than other information channels as a source of reference for consumers. School teachers and the Internet ranked high in research on the marketing of universities. In general, consumers for food products tend to request more information disclosure when using the Internet, such as place of production, whether or not the product is organic, than when using conventional information channels (Yokoyama and Hayashi 2013).

Other research focuses on identifying product groups where the Internet has a comparative advantage as an advertising mode. Watanabe and Iwasaki (2010) studied different products, such as specifics, services or shopping, durable goods, and the reluctance of consumers to purchase online, and found that food ranked as relatively low in consumer reluctance to make online purchases. Katahira (2013) points out that nonessential or luxury products rank low for information exchange online.

Thus, a number of studies have been done on word-of-mouth communication or information exchange online and product groups that are suitable for the medium of the Internet. To date, no research has been published in Japan on the use of the Internet for marketing traditional vegetables as a regional specialty. From a research perspective, this topic is interesting, as these products have the characteristics of food as an essential item, but with one that also has nonessential characteristics. While it may come as a surprise elsewhere, vegetables are frequently purchased as travel souvenirs and gifts in Japan. More research is needed, however, regarding consumer perceptions of regional brands. The collection and analysis of more fundamental data will serve as basis for further developing strategies in the tourism sector.

Objectives and Methods

Hypothesis

Our aim here is to examine if there are any significant differences in the types of advertising media (i.e., Internet, television, personal communication, etc.) between traditional Kaga vegetables.

Our hypothesis: There is a positive relationship between the number of Kaga vegetable items recognized by a person and the types of advertising this person has been exposed to as well as their personal attributes and purchasing amount.

Table 9.2 Number of tourist visits at major sites in Noto area (2012)

Site*	Visits	Site*	Visits
Chirihama Beach	900,000	Wajima Morning Market	644,000
Wakura Onsen (hot spring)	853,000	Notojima Aquarium	417,000
Keta Taisha Shrine	829,000	Aqua Park Shi-On	235,000
Noto Shokusai Market (food market)	713,000	Hyokkori Onsen Shimano Yu (hotspring)	116,000

Sites selected for having over 100,000 annual tourists

Source: Ishikawa Prefecture Tourism Department (2013)

This hypothesis will test the relationship between advertising and recognition of the traditional vegetables in the consumer. The following section elucidates the questionnaire.

The Recognition of Traditional Vegetables by Tourists

Study Sites

There are several sightseeing spots which are associated with foods in Ishikawa. Examples of such sites are illustrated in Table 9.2.

As our focus is food, we selected the Noto Shokusai Market (in Nanao City) and Wajima Morning Market (in Wajima City) which are most closely connected to the food experience in the list above (Fig. 9.1). By choosing these two sites, we focused on respondents with relatively high interest in food. In the Kaga area, the Omi-cho Market in Kanazawa is the equivalent sightseeing spot associated with foods and fish markets.

The characteristics of the two sites are summarized in Table 9.3.

November to December was chosen for the study season, the reason being that Kaga vegetables are produced mainly in winter. In addition, fishery products are available in greater variety during the winter.

A total of 378 tourists were interviewed at the two markets (197 interviewees at the Noto Market, 181 at the Wajima Morning Market). The survey periods were 24 and 25 November and 7 and 8 December 2013. The respondents were all Japanese tourists, and foreign tourists were not included due to the purpose of this study to explore the different media and interactions in Japan.

We carried out face-to-face interviews for the questionnaire, which asked about awareness of traditional vegetable items under the branding categories of Kaga and Noto vegetables.

The items in the questionnaire were transportation mode to the market, purpose of visit, amount spent on purchases, kinds of traditional vegetables respondent

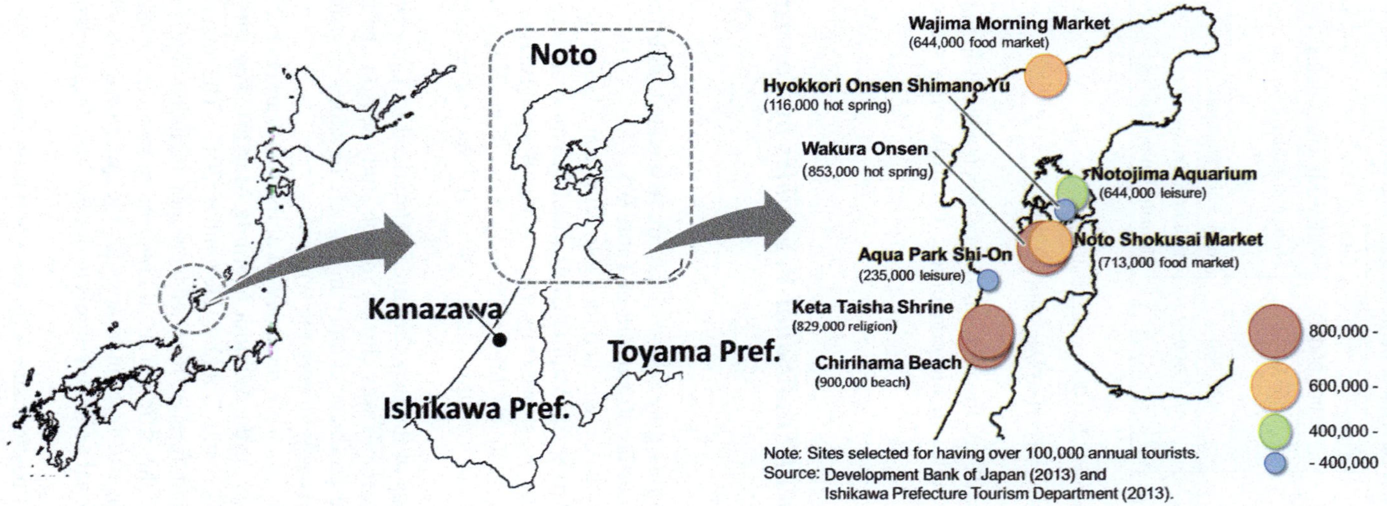

Fig. 9.1 Map of major sites in Noto area (2012)

Table 9.3 Summary of study sites

	Noto Shokusai Market (Nanao fisherman's wharf)	Wajima Morning Market
Location	Nanao City	Wajima City
Annual visitors	713,000	644,000
Features	Roadside station, seaside station, dealing in seafood, vegetable, and processed foods	Dealing in various products including local seafood, vegetables, and Japanese wares

Source: Ishikawa Prefecture Tourism Department (2013)

could identify of each brand, information source about each brand, and personal attributes.

We applied a linear multiple regression model for this analysis. The explanatory variables are set as types of advertisement experienced and personal attributes. The response variables are the number of recognized items of Kaga vegetables (interviewees are ranked 0–15 by the number of items recognized). We gave the parameter x_1 a value of 1 for positive and 0 for negative answers, when asking if the interviewee knew the vegetable for the first time in the shop, in other words whether the shop was the place of encounter with the traditional vegetables. Namely, the parameter of x_1 is given by

$$x_1 = \begin{cases} 1, & |\text{Interviewees knew the vegetables for the first time in the shop} \\ 0, & |\text{Interveiwees did not know the vegetable in the shop} \end{cases}.$$

In a similar way, we gave 1 or 0 to parameter x_2 for the Internet as an information source, parameter x_3 for word-of-mouth communication, x_4 for television, x_5 for newspaper, x_6 for the guidebooks and magazines, x_7 for brochures and posters, and x_8 for others.

For x_9, the value 1 was given for those from outside of prefectures and 0 for those originally from within the prefecture. For amount spent at the markets, x_{10} was used as the parameter. The age groups are categorized by ten-year groupings, with one being for age 10–19 years, two for 20–29 years, and so on, to seven for 70 years and above, for the parameter x_{11}. Moreover, the parameters of x_{10} and x_{11} normalized into [0:1] to compare the difference units.

The parameter x_{12} was gender (male or female). As mentioned, the explanatory variables were mode of advertisement and personal attributes. For personal attributes related to financial elements, we used the amount spent on vegetables. We considered using annual household income, but as some refused to answer, we chose to exclude this parameter.

The linear multiple regression model is expressed by the following formula, with y as the number of Kaga vegetables recognized. With the parameters, the coefficient is determined as follows:

$$y = \sum_{k=1}^{12} \alpha_k x_k + \varepsilon$$

where ε is the intercept. For any parameter with a coefficient of 0, the parameter is not influencing the number of recognized items of Kaga vegetables. We adopted the null hypothesis for the case of $\boldsymbol{\alpha}$ being zero and examined the results. By testing the hypothesis, we assumed that we would be able to test the influence of various advertisement modes.

Results

Through which channels are the consumers gaining their information? We applied the multiple linear regression analysis to 214 datasets from the two markets. The summary of the results is in Table 9.4. With a 95 % of confidence level, the Internet, word-of-mouth communication, travel guidebooks, age of respondents, or brochures and posters are not influencing the purchase amounts. The rest of the items are found to be influencing the behavior of consumers. From the model, female consumers and residents from Ishikawa Prefecture knew a larger number of Kaga vegetable items. The amount they spent did not correlate to the level of knowledge. For the modes of advertising, the Internet as an information source was not a significant channel for consumers.

Table 9.4 Linear regression analysis ($n = 214$)

Variable	Coefficient	*p*-value
Intercept	1.19E+00	0.034815**
Shops on the site	1.92E+00	4.36e−09***
Internet	3.87E−01	0.550757
Oral communication with friends	1.02E+00	0.063433*
Television	1.25E+00	0.000107***
Newspaper	1.03E+00	0.023936**
Guidebooks	−3.94E−02	0.929146
Posters and brochures	1.03E+00	0.109005
Others	2.89E+00	0.000249***
Place of residence	−2.43E+00	3.72e−14***
Amount purchased	4.13E−07	0.988743
Age	1.53E−01	0.052081*
Gender	7.12E−01	0.008306***

The *asterisks* *, **, and *** indicate that the coefficients are statistically different from zero at the 10, 5, and 1 % levels, respectively

The *numbers* in "place of residence" include those who indicated that they are locals in free response portion of the questionnaire

Age was not a significant factor for the number of items recognized. Stronger factors were the experience at shops and television as an advertising mode. The results suggest that the respondents did not depend on the Internet for information about nonessential or luxury foods.

Other factors are relevant to place of residence. In the free comments portion of the questionnaire, number of interviewees mentioned involvement in farming. Examples of other common responses are "Because it is from my homelands," "I am a teacher at local high school in Suzu," and "I live here." Personal attributes such as place of residence seem to have had a greater impact on knowledge of vegetables than the mode of advertising.

Discussion

We have quantified the influence of different modes of advertising on knowledge and purchasing behavior of Kaga traditional vegetable in Japan. Although it is branded as "traditional vegetable," the brand is a relatively new one established in the postwar period.

In other words, we looked quantitatively at the flow of information from producers to consumers. The results indicated that the television, gender, experience at the shops, and place of residence were significant factors. The factor of direct experience at the shop was strong, and this finding has implications for tourism. The consumption of traditional vegetables is compatible with the tourists' expectation for buying local items in person.

The Internet was not influential in promoting purchases, in general. It is too early to conclude clearly about the impact of the Internet, as personal attributes such as place of residence had large impacts. A further exploration is possible whether nonessential items such as traditional vegetables are not well suited to online information exchange. Furthermore, it was observed that tourists expressed their strong preferences to have conversations at the local stores and shops. Such preferences were expressed through the term "direct experience" or "authentic shopping." The tourists prefer to experience and gather information in a rather traditional way, which potentially explains why they regarded the shop on the spot as the main information source.

As for the results of individual items, the Kaga renkon (lotus root) or gorojima kintoki (sweet potatoes) are stable in production volume and well known nationally. If producers aim to establish brands at the national level, it is important to reach out to markets outside the original prefecture and increase production to reach a larger audience.

There are risks with communications as well. If the marketing message is incorrectly transmitted, the cooking methods and original tastes can be lost, which will potentially damage the brand. For example, the Kaga huto kyuri cucumber is cooked and eaten very differently than more common varieties of cucumber. Explanations as to how to cook and eat them will be beneficial for the consumers and for the producers in the long term. Therefore, care is needed to ensure that the flavor and originality are not lost in the process.

Through the questionnaires of tourists at the GIAHS sites of Wajima and Nanao in Noto Peninsula, we identified preliminary trends of the information source of the traditional vegetables. Further researches are necessary to explore whether the case of traditional vegetables is unique or similar to other products or touristic experiences. Further analysis is needed to identify the reason why the information from the Internet did not play a significant role for the consumption of the tourists.

Acknowledgments A number of organizations collaborated in this research. We would like to thank the Noto Vegetable Promotion Association, Suzu City, Noto Town, Noto Shokusai Market and Wajima Morning Market, and interview participants. This study was conducted as a part of the research project MEXT KAKENHI(Grants-in-Aid for Scientific Research 26360062, Project Leader R. Kohsaka; FY2014-FY2016). Part of the project (ES Tradeoff, Project Leader O. Saito, FY2013-FY2015; 1-1303) was also supported by the Ministry of the Environment (MOE) through its Environment Research and Technology Development Fund. The authors' thanks are extended to the valuable discussions at the advisory meeting of Environment Research and Technology Development Fund (S9-1: Project Leader T. Yahara) of the MOE, Japan.

References

Aoya M (2010) Future problem of consumer's behavior in local brand: case study of branding Kyo vegetables (in Japanese). Agric For Res 45(4):343–352. doi:10.7310/arfe.45.343

Buttle FA (1998) Word of mouth: understanding and managing referral marketing. J Strateg Mark 6(3):241–254. doi:10.1080/096525498346658

Dettmann RL, Dimitri C (2009) Who's buying organic vegetables? Demographic characteristics of US consumers. J Food Prod Market 16(1):79–91. doi:10.1080/10454440903415709

Development Bank of Japan (2013) Hokuriku handbook (in Japanese). DBJ, Kanazawa

Development Bank of Japan, Kohsaka Lab (2014) Survey on Kaga vegetables for tourists in Kanazawa. For promotion of Kaga vegetable brand (in Japanese). Retrieved from http://www.dbj.jp/pdf/investigate/area/hokuriku/pdf_all/hokuriku_1403_01.pdf. Accessed 22 Jan 2015

Goto K (2011) Brand communication of purchasing (in Japanese). Univ Mark Distrib Sci Essays: Circ Manag 23(2):25–35, Retrieved from: http://www.umds.ac.jp/kiyou/r/R23-2/025-035.pdf

Hazumi M (2002) Substance of brand theory of primary product—theory discussion of establishment condition and proof survey (in Japanese). Hakuto Shobo Pub., Tokyo

Ishikawa prefectural government Tourism Department (2013) Tourism in Ishikawa prefecture from statistical survey 2012 (in Japanese). Retrieved from: http://toukei.pref.ishikawa.jp/dl/2686/kankoutoukeiH24.pdf

Ishikawa Yasaikaki Research Group (2007) Special product traditional vegetables of Noto, Kaga (in Japanese). Ishikawa Yasaikaki Research Group, Kanazawa

Janssens W, De Pelsmacker P (2005) Advertising for new and existing brands: the impact of media context and type of advertisement. J Mark Commun 11(2):113–128. doi:10.1080/1352726042000306847

Katahira H (2013) No effect on the luxury goods by net communication? (in Japanese). Mark J 33(1):88–98, Retrieved from: http://ci.nii.ac.jp/naid/40019772845/

Kobayashi T (2008) Reliability improvement of the review information based on the bias revision due to a profile and the regionality of the reviewer (in Japanese). The 19th data engineering workshop (DEWS2008) Collection of paper. Retrieved from: http://www.ieice.or.jp/iss/de/DEWS/DEWS2008/proceedings/files/b8/b8-4.pdf

Kotler P (2000) Marketing management, Millennium edition. Prentice-Hall, Inc., Upper Saddle River. Retrieved from: http://www.profit.co.ua/uploads/media/Kotler_Marketing_Management.pdf

Kuribayashi Y (2008) Possibility of university marketing and internet (in Japanese). Nagoya Bunri Univ Bull 8:149–156, Retrieved from: https://nagoya-bunri.ac.jp/information/memoir/2008/2008_20.pdf

Matsushiata R (2007) Each story of Kaga vegetables (in Japanese). Hashimoto-Kakubun-Do, Kanazawa

Nakamura K, Niizawa S (2006) Native district food in the modern eating habits: the actual use situation of Kaga vegetables (in Japanese). Hokuriku Gakuin Short Term Univ Bull 37: 131–154, Retrieved from: http://ci.nii.ac.jp/els/110004782995.pdf

Nishikawa H (2010) Branding and growth of Goroujima-Kintoki (Kaga vegetables) (in Japanese). Hokuriku Crop Sci Soc Rep 45:101–103, Retrieved from: http://ci.nii.ac.jp/els/110004782994.pdf

O'Neill K (2014) Localized food systems? What role does place play? Reg Stud Reg Sci 1(1): 82–87. doi:10.1080/21681376.2014.904596

Otani, N., Saito K (2008) Minor crop and local brand (in Japanese), Continuity and Niche of minor crop management. Economic Research of markets in limited-space. Research result report of technology research expenditure assistance payment, 2. Retrieved from: http://hdl.handle.net/10241/7515

Rath EC (2014) New meanings for old vegetables in Kyoto. Food Cult Soc An Int J Multidiscip Res 17(2):203–223. doi:10.2752/175174414X13871910531944

Saito S (2010) Strategic problem and management system of the local brand (in Japanese). Agric For Bus Res 45(4):324–335. doi:10.7310/arfe.45.324

Saito S (2011) Food industry cluster and local brand in Japan (in Japanese). Food Sys Res 17(2): 90–96. doi:10.5874/jfsr.17.90

Sato Y, Nishikawa Y, Tsurubuchi T (2012) Consideration about the local promotion of the roadside station (in Japanese). Nagasaki Wesleyan Univ Mod Soc Dept Bull 10(1):53–62, Retrieved from: http://ci.nii.ac.jp/els/110009323577.pdf

Savitri D (2008) An approach of sustainable development: rural revitalization as the pioneer of OVOP movement. J OVOP Policy 7:21–30, Retrieved from: http://www.iovoppa.org/journal/01_200810/en/07_dyah-e.pdf

Watanabe K, Iwasaki K (2010) Product type based on the consumer awareness in the net purchasing. Tokyo City Univ Environ Inf Dept Bull 11:1–9, Retrieved from: http://www.yc.tcu.ac.jp/~cisj/11/11_24.pdf

Yamamoto M (2004) Review of Kaga vegetables. Hokuriku Econ Res 313:1–15

Yokoyama K, Hayashi H (2013) Effects of disclosure of production history information in agricultural products on buying motivation (in Japanese). Tsukuba J Agric For 1:19–29. Retrieved from: http://www.nourin.tsukuba.ac.jp/tjaf/pdf/nourin-g2013_19-29.pdf

Index

S.-i. Nakano et al. (eds.), *Aquatic Biodiversity Conservation and Ecosystem Services*, Ecological Research Monographs, DOI 10.1007/978-981-10-0780-4

The manufacturer's authorised representative in the EU is Springer Nature Customer Service Centre GmbH, Europaplatz 3, 69115 Heidelberg, Germany. If you have any concerns regarding our products, please contact ProductSafety@springernature.com

Printed and bound by CPI Group (UK) Ltd, Croydon, CR0 4YY
28/09/2026
02224085-0016